ENERGY METABOLISM

ENERGY METABOLISM

Eric G. Ball

Marine Biological Laboratory
Woods Hole, Massachusetts

1973
ADDISON-WESLEY PUBLISHING COMPANY, INC.
ADVANCED BOOK PROGRAM
Reading, Massachusetts

London • Amsterdam • Don Mills, Ontario • Sydney • Tokyo

Library of Congress Cataloging in Publication Data

Ball, Eric Glendinning.
Energy metabolism.

Bibliography: p.
1. Energy metabolism. I. Title.
QP171.B32 574.1'33 72-4389
ISBN 0-201-00406-2
ISBN 0-201-00407-0 (pbk.)

Printed in the United States of America

ISBN 0-201-00406-2 (hardbound)
ISBN 0-201-00407-0 (paperback)
ABCDEFGHIJ-MA-79876543

Dedicated to my students,
medical and graduate,
who taught me how to study

CONTENTS

LIST OF FIGURES

PREFACE

This book is not written for the expert in the field, but for the person who wishes to obtain a grasp of the basic principles and reactions underlying the conversion of foodstuffs into energy. It is the outgrowth of a series of lectures given to first-year medical students in which an attempt was made to put together a coherent story of the various aspects of energy metabolism. Though these reactions are dealt with in nearly all textbooks of biochemistry, they are usually so dispersed throughout the book that the student loses the thread of the total story. In this volume the basic reactions involved in biological oxidations at the substrate level are given first. For the sake of simplicity in presentation, these reactions are divided into three categories: dehydrogenations, decarboxylations and "make-ready." The latter term is applied to all reactions in which alterations in a metabolite occur that lead to a product that can undergo dehydrogenation and/or decarboxylation. The reactions involved in the conversion of glucose and fatty acid to CO_2, H^+, and electrons are presented from this viewpoint, with the acceptance of the electrons and some of the H^+ by those coenzymes that are termed primary electron acceptors. The passage of electrons from these primary electron acceptors to oxygen over the electron transmitter system and the generation of high-energy phosphate then follows.

A brief chapter is devoted to the localization within the cell of the reactions involved in biological oxidations, and to some of the problems posed by membrane barriers. A chapter entitled "Energy Storage" deals with the reactions involved in the conversion of glucose to glycogen and fatty acids. In this chapter the pathways for the formation of TPNH used in fatty acid synthesis are presented. A final chapter deals with energy mobilization and glucogenesis. In the reactions involved in the breakdown of glycogen and fat to supply the energy needs of the body, the role of hormones and cyclic AMP in these mobilization processes is stressed.

CHAPTER 1

Biological Oxidations and the Basic Pattern of the Process

The energy demands of the living organism are met mainly by the combustion of glucose and fatty acids. The equations that depict the overall reactions of these processes are

	R.Q.	Kcal/g
$C_6H_{12}O_6 + 6O_2 \longrightarrow 6CO_2 + 6H_2O$	1.0	4.1
$C_{16}H_{32}O_6 + 23O_2 \longrightarrow 16CO_2 + 16H_2O$	0.7	9.5

where palmitic acid is used as a representative of fatty acids. In each of these reactions energy is released, energy which originally had its source in the sun and which was trapped in these foodstuffs as chemical energy by the photosynthetic reactions of plants. The energy released on the combustion of these substances is expressed above in terms of kilocalories per gram generated as heat. Though these values are used in expressing the energy needs of the human body for dietary purposes, they are not exact from the standpoint of the energy yield to the animal. This is because the living cell is not a heat engine and can make use only of the free energy released in these reactions. However, the heat production values are close approximations of the free energy release and serve to indicate that fatty acids, per unit weight, are a far better source of energy than glucose. The ratio of CO_2 produced to the oxygen consumed in these two reactions, the respiratory quotient (R.Q.), differs. This fact permits one by measurement of the O_2 consumption and CO_2 production of the living organism to estimate the relative amounts of the two substances that are undergoing oxidation and hence the calories or energy consumption under any given metabolic state.

The overall equations given above for the oxidation of glucose and palmitic acid might be interpreted to indicate that oxygen reacts directly with these substances to yield CO_2 and H_2O. In the living cell this is not the case. In essence the process involves the transfer of electrons from the foodstuff to oxygen in a stepwise fashion over a chain of compounds. Each participant in this chain constitutes an oxidation–reduction system; it is oxidized by the removal of electrons and is reduced when it accepts electrons. The simplest form of such a system is that represented by the interconversion of ferrous and ferric ions:

$$Fe^{++}(X) \rightleftarrows Fe^{3+}(X) + e$$

Here (X) represents a substance with which the iron may be chelated, and *e* represents an electron which must be accepted or donated by another substance for the reaction to proceed. Since one electron is involved, it is termed a one-electron system. A second and more common type of oxidation–reduction reaction in biological processes is that represented by the oxidation of hydroquinone to quinone:

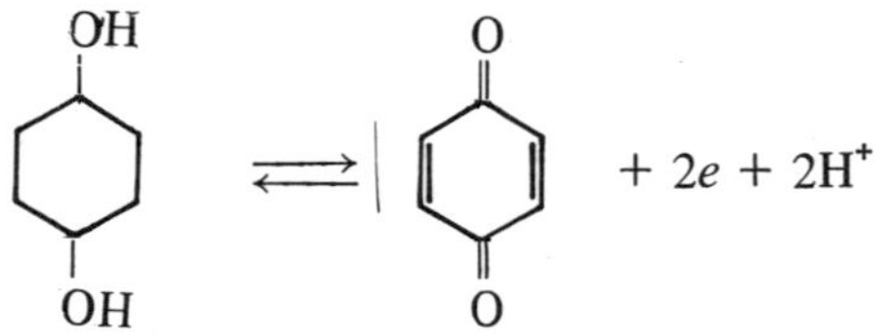

In this case two electrons and two hydrogen ions are removed during the oxidation, and the system is termed a two-electron system. Each oxidation–reduction system is characterized by an oxidation–reduction potential which is expressed as the voltage difference at a fixed temperature between a solution containing equal amounts of the oxidant and reductant of the system and some agreed-upon standard oxidation–reduction system. The pH of the solution must be specified when H^+ is involved in the reaction, as, for example, the hydroquinone–quinone system, since the potential of such systems will vary with pH. In the case of $Fe^{++} - Fe^{3+}$ systems, which play an important role in biological oxidation, it should be noted that the potential of the system will vary markedly depending on the nature of (X).

The extent and direction of interaction between two oxidation–reduction systems can be predicted if their potentials are known. For example, let us suppose that we are dealing with the ferri–ferrocyanide system ($X = CN^-$). At pH 7.0 and 30°C the potential of this system is +0.425 volt as measured against the standard system, H_2 (1 atm) : H^+ (1.0 normal). The hydroquinone–quinone system under the same condition yields a potential of +0.275 volt. We can write then the following equation:

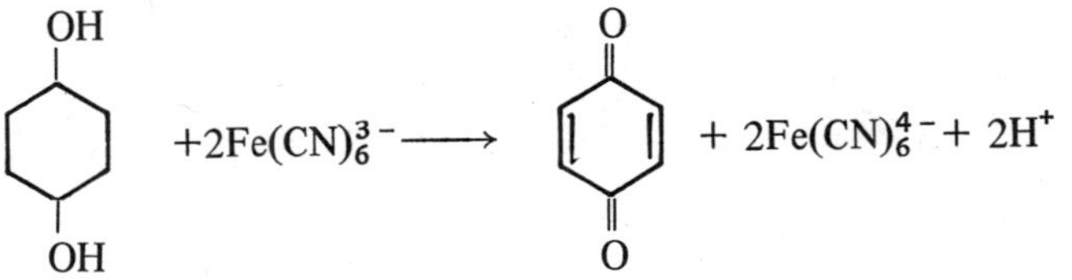

Since the potential of the ferro–ferricyanide system is positive to the hydroquinone–quinone system, we can predict that ferricyanide will remove electrons from hydroquinone and the reaction will proceed in the direction of the arrow. If 1 mole of hydroquinone is mixed with 2 moles of ferricyanide, we can also predict that the potential of the resulting mixture will be halfway between that of the two systems reacting, or +0.350 volt and that over 99% of the two starting materials will have undergone an exchange of electrons.

In examining the series of oxidation–reduction reactions that occur in biological oxidation, it is convenient for the purposes of our presentation to classify them as shown in Fig. 1-1. The first step is the removal of electrons and hydrogen ions from the metabolite and the acceptance of the electrons by what will be termed a primary electron acceptor. As shown in the diagram, it is at this step that all the CO_2 produced in biological oxidations is formed. A small amount of the total energy to be released in the overall pathway may be converted into high-energy phosphate ($\sim$P) at this step. This process will be referred to as phosphorylation at the substrate level. The second step in this presentation will be the passage of electrons from the primary electron acceptor into the chain of oxidation–reduction systems constituting the electron transmitter system (E.T.S.) The hydrogen ions removed simultaneously from the substrate with the electrons do not follow hand in hand with the passage of electrons over the E.T.S. to oxygen. In some cases for each electron transferred a hydrogen ion is simultaneously taken up, in others only one hydrogen ion for each two electrons, and in others none. An attempt is made to represent this fact in the diagram by the dotted lines. The bulk of the high-energy phosphate formed in the overall process is produced in step II. In this process, referred to as oxidative phosphorylation, energy equivalent to 8000 to 9000 calories and provided by an oxidation–reduction reaction is captured by the conversion of adenosine diphosphate (ADP) and inorganic phosphate (P_i) to adenosine triphosphate (ATP):

$$\text{ADP} + \text{P}_i \longrightarrow \text{ATP}$$

Adenosine triphosphate has the following structure:

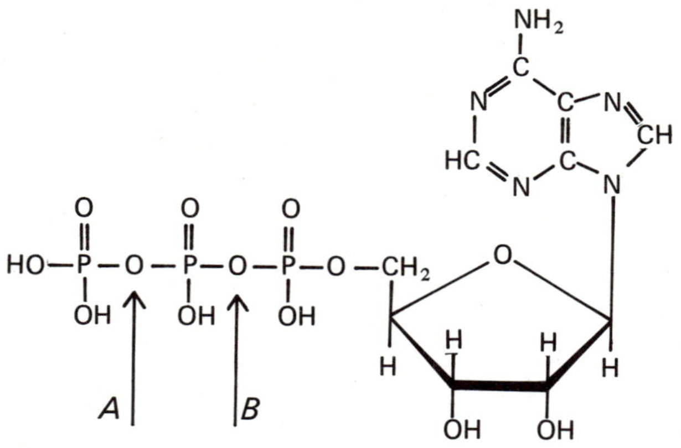

Adenosine Triphosphate (ATP)
Hydrolysis at *A* or *B* yields 8000 to 9000 calories per mole (high-energy phosphate bond). Hydrolysis at *A* produces inorganic phosphate (P_i) and adenosine diphosphate (ADP). Hydrolysis at *B* yields pyrophosphate (PP_i) and adenosine monophosphate (AMP) (adenylic acid).

It is to be noted that all the oxygen consumed is reduced to water. At rest a human being of average size consumes approximatley 250 ml of oxygen per minute. This amount of oxygen will yield about 25 ml of H_2O per hour, or 600 ml/day. Not all this water is net gain, since, as will be seen later, some water is consumed by the metabolic pathways involved in its production. In some species this water formed by oxidative processes plays a major role in the economy of the animal. For example, birds excrete very little water as compared to man and need a much smaller water intake to supplement the water produced by oxidative processes and that consumed with their food. Animals that inhabit the desert also rely heavily on water generated by metabolic processes.

In later chapters there will be presented the specific reactions by which substrates (glucose, fatty acids, and their metabolites) donate electrons to a primary electron acceptor and produce CO_2. Before considering these reactions it will be helpful in understanding them to point out certain basic principles which will be encountered. For this purpose it is convenient to divide these reactions into three categories: (1) dehydrogenation, (2) decarboxylation, and (3) make-ready.

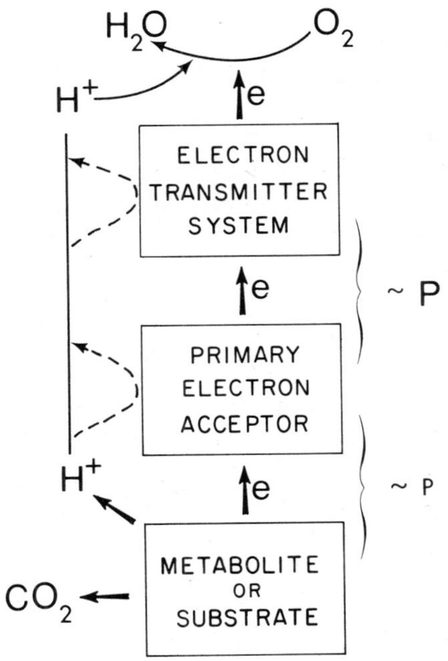

FIG. 1-1. The basic pattern of biological oxidations. The overall process depicted here is sometimes called the respiratory chain. This term and the term electron transport system are also used to describe the chain of reactions encompassing the primary electron acceptors and what herein is called the electron transmitter system. The word transmitter is employed to describe this system because its components, unlike the primary electron acceptors, are not in true solution and hence the passage of electrons through this system resembles to some extent the passage of electrons over a wire.

Dehydrogenation

All oxidations at the substrate level may be called dehydrogenations, since they involve the removal of two hydrogen atoms or, what is the equivalent, two hydrogen ions and two electrons. Two main types of chemical grouping are discernible in the dehydrogenation of substrates. First is the grouping –CHOH–; as an example we may use the dehydrogenation of lactate to form pyruvate:

$$\mathrm{CH_3\cdot\underline{CHOH}\cdot COO^-} \rightleftarrows \mathrm{CH_3\cdot\underline{CO}\cdot COO^-} + 2\mathrm{H^+} + 2e$$

The reaction is written here as a half-reaction, and of course it will not proceed unless a substance, here called the primary electron acceptor, is available to accept the electrons. As we shall subsequently see, the hydrogen ions may be also passed on to this acceptor or released into the environment. An enzyme must be present to catalyze the reaction; enzymes of this type are called dehydrogenases and are named for the substrate from which hydrogen ions and electrons are removed. In the example given, this would be lactate dehydrogenase. It should be noted that the reaction is reversible and that the enzyme also catalyzes the reduction of pyruvate and thus might be also called pyruvate reductase. The second type of chemical grouping frequently encountered in biological dehydrogenations is $-CH_2\cdot CH_2-$. As an example of this reaction we may use the conversion of succinate to fumarate:

$$\mathrm{COO^-\cdot\underline{CH_2\cdot CH_2}\cdot COO^-} \rightleftarrows \mathrm{COO^-\cdot\underline{CH = CH}\cdot COO^-} + 2\mathrm{H^+} + 2e$$

The main point to be emphasized is that the oxidation of all substrates always involves the removal of two hydrogen ions and two electrons.

Decarboxylation

All the CO_2 produced in biological oxidations is formed by a process called decarboxylation. This may be represented in its simplest terms as the conversion of a carboxyl group to CO_2 in the following manner, where R represents the rest of the substrate:

$$\mathrm{R\cdot CO\cdot COOH} \longrightarrow \mathrm{R\cdot C}\genfrac{}{}{0pt}{}{\nearrow\!\mathrm{O}}{\mathrm{H}} + \mathrm{CO_2}$$

In all the reactions to be considered later this process occurs *pari passu* with a dehydrogenation reaction, and a stepwise separation of the two processes on an experimental basis is difficult. No specific enzyme other than the dehydrogenase may be needed for the decarboxylation reaction to occur, though the presence of metal ion may be essential. As will be seen later, the actual production of a free aldehyde group, $\mathrm{R\cdot C}\genfrac{}{}{0pt}{}{\nearrow\!\mathrm{O}}{\mathrm{H}}$, does not take place.

Make-Ready

This term has been coined as a convenient means to designate a reaction or a series of reactions in which a substrate undergoes alteration in preparation for a dehydrogenation. In many cases such reactions are a means for producing in a substrate either a –CHOH– or a $-CH_2 \cdot CH_2-$ grouping where none existed before. The simplest example of such a reaction is the addition of H_2O. For example, the fumarate produced by the dehydrogenation of succinate in the reaction given above can be made ready for another dehydrogenation by the addition of water, namely:

$$COO^- \cdot CH = CH \cdot COO^- + H_2O \rightleftharpoons COO^- \cdot \underline{CHOH} \cdot CH_2 \cdot COO^-$$

There is thus formed the grouping –CHOH–, which is susceptible to a dehydrogenation step. A specific enzyme is required for such a process, and in the case of fumarate this enzyme is called fumarase.

Primary Electron Acceptors

The acceptors of electrons from substrates fall mainly into two types – the so-called pyridine nucleotides and the flavin nucleotides. As a general rule, a pyridine nucleotide acts as the primary electron acceptor when the substrate grouping undergoing oxidation is –HCOH–. When the grouping undergoing dehydrogenation is $-CH_2 \cdot CH_2-$, then a flavin nucleotide is usually involved as a primary acceptor.

Two pyridine nucleotides are known which function as primary electron acceptors. Diphosphopyridine nucleotide (DPN^+), also called nicotinamide adenine dinucleotide (NAD), is usually employed as an electron acceptor when the electrons are to be subsequently handed over to the electron transmitter system for the generation of high-energy phosphate. The other, triphosphopyridine nucleotide (TPN^+), also called nicotinamide adenine dinucleotide phosphate (NADP), is largely employed when substrate electrons are to be used for synthetic purposes. An example of this use will be encountered when the conversion of glucose to fatty acids is considered.

The part of these coenzymes that functions in the acceptance of electrons is the nicotinamide portion which is derived from the vitamin nicotinic acid. The reaction may be written in the following manner:

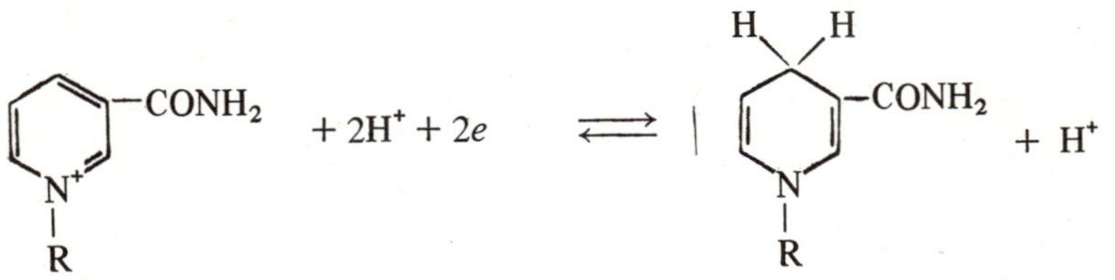

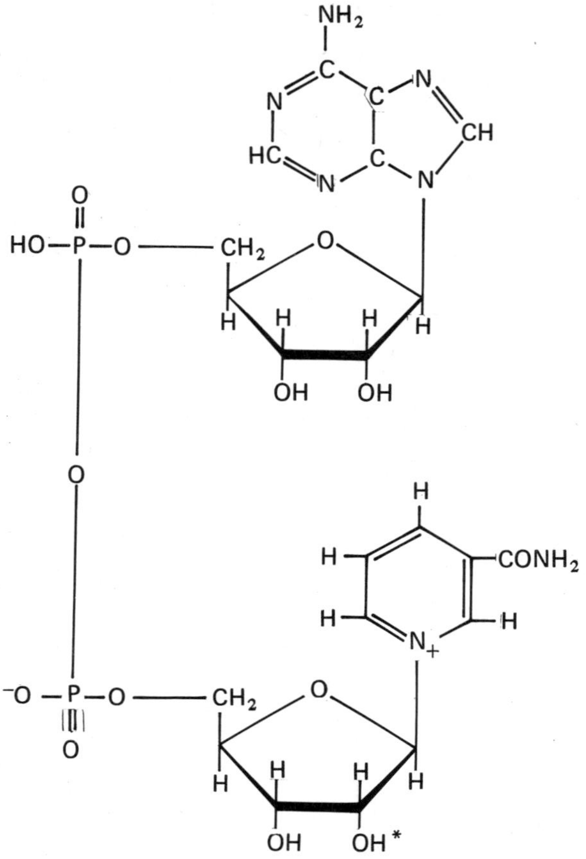

Diphosphopyridine Nucleotide (DPN^+)
Triphosphopyridine nucleotide (TPN^+) is formed when the OH^* group is replaced by $—OPO_3H^-$.

The reaction is written in this form to emphasize the fact that, though both electrons are accepted, only one of the two hydrogen ions removed from the substrate becomes attached to form the reduced pyridine nucleotide. One of the hydrogen ions is released and temporarily at least must be absorbed by the buffering capacity of the cell. Hence the reduced forms of the pyridine nucleotides are represented by the abbreviations DPNH and TPNH, or NADH and NADPH. On reduction, the pyridine nucleotides develop an absorption band centered at 340 mμ. The appearance or disappearance of this band may be used to follow quantitatively the participation of these compounds in dehydrogenase-catalyzed reactions. It should be noted that, in general, DPN^+ cannot effectively substitute for TPN^+, and vice versa, as a primary electron acceptor in a specific dehydrogenase reaction. The additional phosphate grouping on the adenylic acid portion of TPN^+ would appear to be the determining factor in this marked specificity.

The other type of primary electron acceptor, the flavin nucleotides, are usually participants in a reaction when a $-CH_2-CH_2-$ grouping in the substrate is to be dehydrogenated. Two flavin nucleotides are known – flavin mononucleotide (FMN) and flavin adenine dinucleotide (FAD).

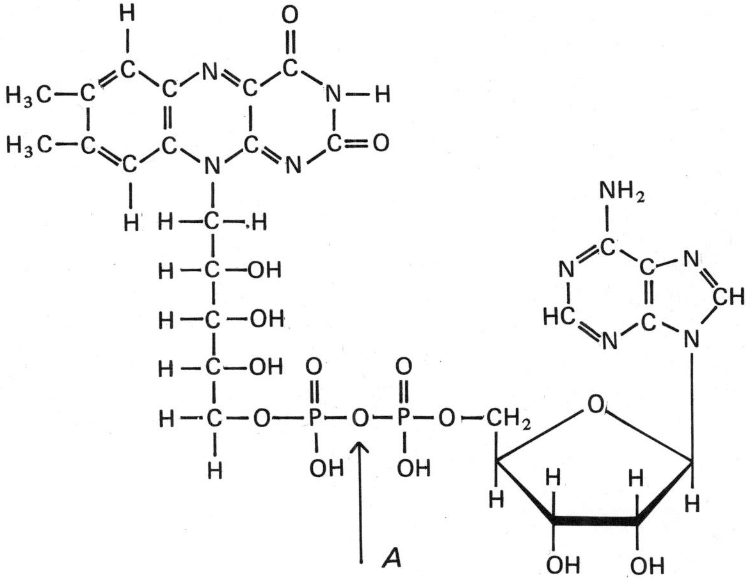

Flavin Adenine Dinucleotide (FAD)
Hydrolysis at *A* yields flavin mononucleotide (FMN) and adenylic acid (AMP).

The flavin nucleotides contain the vitamin riboflavin, and it is the isoalloxazine ring of this vitamin which acts as the electron acceptor in the following manner:

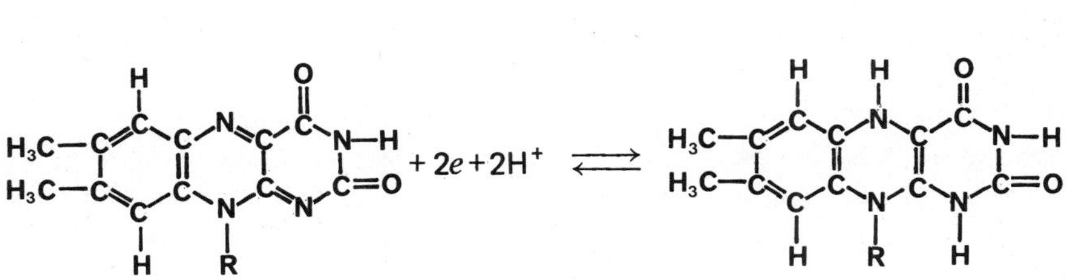

Here two electrons and two hydrogen ions are accepted by the yellow oxidant to form the reduced colorless form. As in the case of the pyridine nucleotide, the reaction therefore may be measured spectrophotometrically.

The pyridine and flavin nucleotides in most biological oxidations accept electrons and hydrogen ions directly from the substrate. An exception occurs when the substrate is an α-keto acid – for example, pyruvate or ketoglutarate. Here a sulfhydryl compound, lipoic acid, acts as intermediate between the substrate and what we have called the primary electron acceptors. This situation will be dealt with in more detail later.

The reaction between the primary electron acceptors and the substrate occurs only in the presence of the dehydrogenase that is specific for the substrate. In the case of a pyridine nucleotide there is generally no firm attachment between it and the dehydrogenase. We shall therefore write such reactions, for example, as follows:

$$CH_3CHOH{\cdot}COOH + DPN^+ \rightleftharpoons CH_3CO{\cdot}COOH + DPNH + H^+$$

it being understood that a specific dehydrogenase participates in the reaction. The DPNH (or TPNH) formed is thus capable of existing free in solution and may be utilized in conjunction with another dehydrogenase to effect a reduction. The situation is different for the flavin nucleotides. Here a firm attachment usually exists between the flavin nucleotide and the protein that catalyzes the specific reaction. We shall therefore write reactions of this type as involving a flavoprotein (FP). For example in the reaction

$$COOH{\cdot}CH_2{\cdot}CH_2{\cdot}COOH + FP \rightleftharpoons COOH{\cdot}CH = CH{\cdot}COOH + H_2FP$$

FP is succinate dehydrogenase and represents both the enzyme and the electron acceptor. Unlike DPNH, it is not possible for H_2FP to react as a reductant for a number of metabolic intermediates.

Direct oxidation of a reduced pyridine nucleotide by oxygen occurs very slowly at best and is not of importance in biological processes. Some reduced flavoproteins resemble DPNH or TPNH in that they react poorly with oxygen and are therefore to be classed as dehydrogenases – for example, succinate dehydrogenase. Other reduced flavoproteins react rapidly with oxygen, and to distinguish them from dehydrogenases they are called oxidases – for example, xanthine oxidase, glucose oxidase, and D-amino acid oxidase. However, as in a dehydrogenase reaction, two electrons and two hydrogen ions are removed from the substrate by the oxidase to form H_2FP. In the oxidase reaction the electrons and hydrogen ions are handed on directly to O_2 to form H_2O_2 instead of forming H_2O as is the case when they react with O_2 through the mediation of the electron transmitter system. The energy released in an oxidase reaction therefore is presumably wasted, since high-energy phosphate production appears to occur only when the flow of electrons to O_2 is by way of the electron transmitter system.

The amount of oxygen consumed in oxidase reactions is not a significant percent of the total O_2 consumption of living cells. Of more importance is the fact that the H_2O_2 generated can be detrimental to the cell if allowed to accumulate. The cell has two means for the disposal of H_2O_2. First, the enzyme catalase accelerates the breakdown of H_2O_2 to water and oxygen.

$$2H_2O_2 \longrightarrow 2H_2O + O_2$$

Second, peroxidases catalyze reactions in which two electrons and two hydrogen ions are removed from a substrate and are made to react with H_2O_2 to produce two molecules of water.

$$R{\cdot}H_2 + H_2O_2 \longrightarrow R + 2H_2O$$

Both catalase and peroxidase are enzymes that contain iron phorphyrin groupings.

CHAPTER 2

Fatty Acid Oxidation: Citric Acid Cycle

The bloodsteam transports both glucose and fatty acids to the tissues to supply their energy requirements. On entering the tissue both substances must be made ready for dehydrogenation by processes that require the expenditure of energy. It may be said that it is necessary to prime the pump in order to activate the process by which energy release from these substances is achieved. Hence reactions of this type will be included under the category "make-ready." Activation of both glucose and fatty acid is accomplished by reactions involving ATP in which either a phosphate group or adenylic acid from ATP becomes linked to the substrate. Enzymes that catalyze reactions of this type are termed kinases. Since the reactions for the initial steps in fatty acid oxidation are the simplest, we shall take them up first. They fall into the two categories of reactions which we have chosen to label make-ready and dehydrogenation.

The preparation of fatty acids for oxidation involves the use of coenzyme A. This coenzyme is composed of β-mercaptoethylamine (*A*), attached in amide linkage to the vitamin pantothenic acid (*B*), which in turn is bound in ester linkage to adenosine 3′-phosphate 5′-pyrophosphate (*C*). The abbreviation CoA•SH or CoA will be used for coenzyme A.

The reactions involved in the oxidation of fatty acids to acetyl CoA are given in Fig. 2-1. Reaction 1, classified as a make-ready reaction, involves the formation of a thioester linkage between the fatty acid carboxyl group and the SH group of coenzyme A. In the first step the linkage between the fatty acid and AMP involves the phosphate group of adenylic acid and forms the high-energy bond $-\overset{O}{\overset{/\!/}{C}}\sim O\cdot\overset{O}{\overset{\|}{\underset{OH}{\underset{|}{P}}}}-O-$. The pyrophosphate ($PP_i$) which is produced undergoes hydrolysis to inorganic phosphate by the action of a pyrophosphatase. In the second step the SH group of coenzyme A replaces adenylic acid in the ester linkage to form a thioester, $-\overset{O}{\overset{/\!/}{C}}\sim S-\overset{H_2}{\overset{|}{C}}-$ which is also a high-energy bonding. Thus in the overall reaction of the two high-energy bonds in ATP, one is lost through the formation of pyrophosphate and the other is retained in the thioester linkage. One enzyme, termed a thiokinase, is responsible for catalyzing the overall reaction. A number of thiokinases are known which vary in activity according to the number of carbon atoms in the fatty acid used as substrate and are named accordingly. Thus acetic thiokinase acts on C_2 and C_3 fatty acids, octanoic thiokinase on C_4 to C_{12} fatty acids, and dodecanoic thiokinase on C_{10} to C_{18} fatty acids.

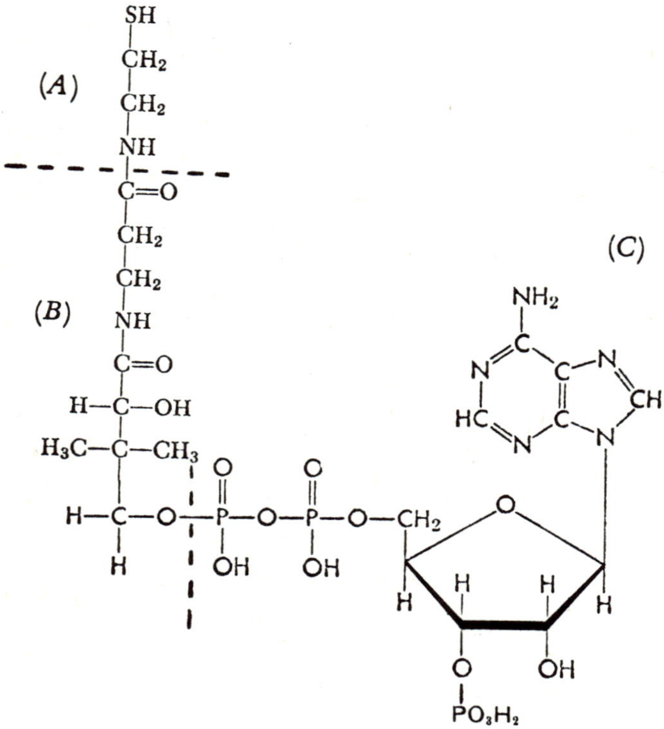

Coenzyme A
Hydrolysis at the dotted lines yields (*A*) β-mercaptoethylamine, (*B*) pantothenic acid, and (*C*) adenosine 3′-phosphate 5′-pyrophosphate.

The CoA derivative of a fatty acid is capable of undergoing a dehydrogenation as shown in reaction 2, Fig. 2-1. Since this dehydrogenation involves the removal of two electrons and two hydrogen ions from a $-CH_2-CH_2-$ group to form a $-CH{=}CH-$ group, the primary electron acceptor employed is a flavoprotein, which is called an acyl CoA dehydrogenase. Here, as in the case of the thiokinases, a number of acyl CoA dehydrogenases are known which are

FATTY ACID OXIDATION

1. $R{\cdot}CH_2{\cdot}CH_2{\cdot}COOH + ATP \xrightarrow[\text{Thiokinase}]{} R{\cdot}CH_2{\cdot}CH_2{\cdot}CO{\cdot}AMP + PyroPO_4$

$R{\cdot}CH_2{\cdot}CH_2{\cdot}CO{\cdot}AMP + HSCoA \longrightarrow R{\cdot}CH_2{\cdot}CH_2{\cdot}COSCoA + AMP$

2. $\underline{CH_3(CH_2)_{10}CH_2{\cdot}CH_2{\cdot}CH_2{\cdot}CH_2{\cdot}COSCoA} + FP \underset{\text{Acyl dehydrogenase}}{\rightleftharpoons} CH_3(CH_2)_{10}CH_2{\cdot}CH_2{\cdot}CH = CHCOSCoA + H_2FP$

3. $CH_3(CH_2)_{10}CH_2{\cdot}CH_2{\cdot}CH = CHCOSCoA + H_2O \underset{\text{Enoyl hydrase}}{\rightleftharpoons} CH_3(CH_2)_{10}CH_2{\cdot}CH_2{\cdot}CHOH{\cdot}CH_2{\cdot}COSCoA$

4. $\underline{CH_3(CH_2)_{10}CH_2{\cdot}CH_2{\cdot}CHOH{\cdot}CH_2{\cdot}COSCoA} + DPN^+ \underset{\beta\text{-Hydroxyacyl dehydrogenase}}{\rightleftharpoons} CH_3(CH_2)_{10}CH_2{\cdot}CH_2{\cdot}CO{\cdot}CH_2{\cdot}COSCoA + DPNH + H^+$

5. $CH_3(CH_2)_{10}CH_2{\cdot}CH_2{\cdot}COCH_2{\cdot}COSCoA + HSCoA \underset{\beta\text{-Ketoacylthiolase}}{\rightleftharpoons} \underbrace{CH_3(CH_2)_{10}CH_2{\cdot}CH_2{\cdot}COSCoA}_{\downarrow} + \underset{\text{Acetyl CoA}}{CH_3COSCoA}$

To Reaction 2 and repeat sequence

6. $C_{16}H_{32}O_2 \underset{?}{\rightleftharpoons} 8\,CH_3COSCoA + \underbrace{7\,H_2FP + 7\,DPNH}_{28\,e} + 7H^+$

FIG. 2-1. The oxidation of fatty acids to acetyl CoA.

named to indicate the carbon chain length of the acyl group most readily attacked – butyryl, octanoyl, and hexadecanoyl CoA dehydrogenases. The next step in the sequence of reactions, reaction 3, Fig. 2-1, is a make-ready reaction in which H_2O is added at the double bond formed by the dehydrogenation reaction. The enzyme for the reaction is named enoyl CoA hydrase and appears to act on substrates of all chain lengths. The hydration reaction results in the production of a –CHOH– grouping, and hence the product becomes ready for another dehydrogenation step, reaction 4, Fig. 2-1, which employs DPN^+ as the primary electron acceptor. The enzyme for this reaction is named β-hydroxyacyl dehydrogenase, and one enzyme appears to serve for substrates of different chain lengths. The final step in the cycle is a thiolase reaction, reaction 5, in which CoA participates to yield acetyl CoA and an acyl CoA with two less fatty acid carbons than the starting material. Several thiolases exist which possess different chain length specificity. Since the acyl CoA produced is now ready for another dehydrogenation, the reaction may be termed a make-ready reaction.

The sequence of reactions 2 through 5 may now be repeated as many times as is necessary to convert all the fatty acid starting material into acetyl CoA. In the case of the C_{16} saturated fatty acid, palmitic, used as an example in Fig. 2-1, a total of seven cycles would be required. The result, as shown in reaction 6, Fig. 2-1, would be the formation of 8 acetyl CoA, 7 H_2FP, 7 DPNH, and 7 H^+. It will be noticed that neither CO_2 nor high-energy phosphate is produced in this series of reactions. It is understood, of course, that in the cell the products of the reaction do not accumulate, but for the purposes of this presentation we shall assume that they do.

The Citric Acid Cycle

The acetyl CoA produced by the oxidation of fatty acids is fed into the citric acid cycle where it undergoes further dehydrogenation and complete conversion to CO_2. In presenting the citric acid cycle it may be helpful if we accept the premise that the cell is limited in its dehydrogenation machinery to the removal of two electrons and two hydrogen ions from the two groupings –CHOH– and $–CH_2 \cdot CH_2–$. The citric acid cycle may then be looked upon as an ingenious device contrived by the cell to convert CH_3COOH into forms in which these two groupings appear. At any rate the cell achieves just this by combining acetic acid with oxalacetate, a four-carbon compound, to produce citric acid, a six-carbon compound. There then follows a series of reactions with the formation of compounds containing either a –CHOH– or a $–CH_2 \cdot CH_2–$ grouping which are submitted to dehydrogenation reactions. In the end, oxalacetate is re-formed, and the equivalent of one molecule of acetic acid is converted to CO_2.

The citric acid cycle is portrayed in Fig. 2-2. For the sake of simplicity, in this diagram the carboxyl groups of all compounds are shown as un-ionized, though at the pH of the cell they will be mainly in the ionized form. The initial step, labeled 1 in the figure, is a make-ready reaction involving the condensation of acetyl CoA with oxalacetate to form citrate. The reaction is catalyzed by what is called "condensing enzyme," or citrate synthetase. The equilibrium for this reaction is such that it can be said to be irreversible in the cell. This is understandable when it is remembered that the thioester linkage in acetyl CoA is a high-energy bond. Indeed, as we shall see later, the cleavage of citrate to acetyl CoA and oxalacetate does occur in the cell, but it is a reaction that requires the utilization of ATP and another enzyme, citrate lyase.

If one looks at the structure of citric acid, one sees that it does not contain either a $–CH_2 \cdot CH_2–$ or a –CHOH– grouping. The cell therefore carries out another make-ready reaction which produces isocitric acid which contains a –CHOH– grouping. This is accomplished by a reaction (2 and 3), which can be visualized as the dehydration of citric acid to form *cis*-aconitic acid followed by the readdition of water across the double bond that is formed to produce isocitric acid. The enzyme catalyzing this reaction is named aconitase and displays maximum activity *in vitro* in the presence of Fe^{++} and a sulfhydryl compound – for example, glutathione or cysteine. A stereospecific attachment of citric acid to the enzyme causes the rearrangement to occur only in the portion of the citrate molecule that is derived from oxalacetate. As a result, in one turn of the citric acid cycle, none of the carbon atoms derived from acetyl CoA appear in the two CO_2 molecules produced.

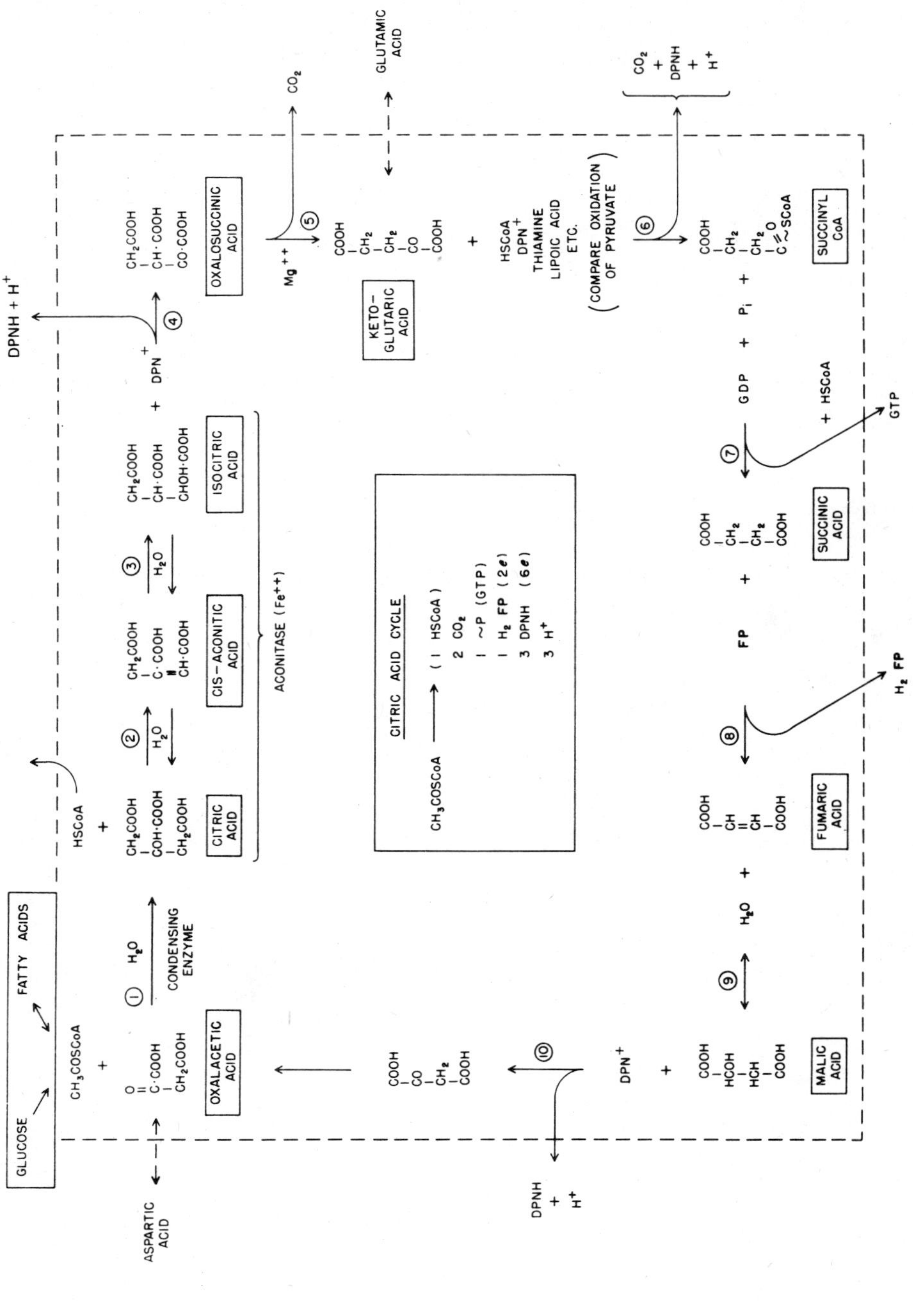

FIG. 2-2. The citric acid cycle.

Since isocitric acid contains the grouping –CHOH–, it can undergo a dehydrogenation in which DPN^+ is the primary electron acceptor (4). As rapidly as the oxidation product, oxalosuccinic acid, is formed, it undergoes a decarboxylation reaction (5), forming α-ketoglutaric acid. The decarboxylation step is catalyzed by the same enzyme, isocitrate dehydrogenase, which carries out the dehydrogenation. Magnesium ion attached to the enzyme appears to facilitate the decarboxylation reaction. This step then represents the first production of CO_2 from the fatty acid starting material.

The next step (6) in the citric acid cycle, the oxidation and decarboxylation of an α-keto acid, is the most complex of all that are to be encountered in biological oxidations. Since it will be met with again when we consider the metabolism of pyruvate formed in the oxidation of glucose, it will be presented here using R•CO•COOH to represent either α-ketoglutaric acid or pyruvic acid. It should be noted that a similar but distinct enzyme complex is required for each of these acids.

The multiple reactions that occur in the decarboxylation and dehydrogenation of an α-keto acid take place on the surface of a multienzyme complex with a molecular weight in the millions. Three cofactors – thiamine pyrophosphate, lipoic acid, and flavin adenine dinucleotide – are attached to this complex and participate in the reactions. The initial event is a decarboxylation in which the pyrophosphate ester of the vitamin thiamine participates.

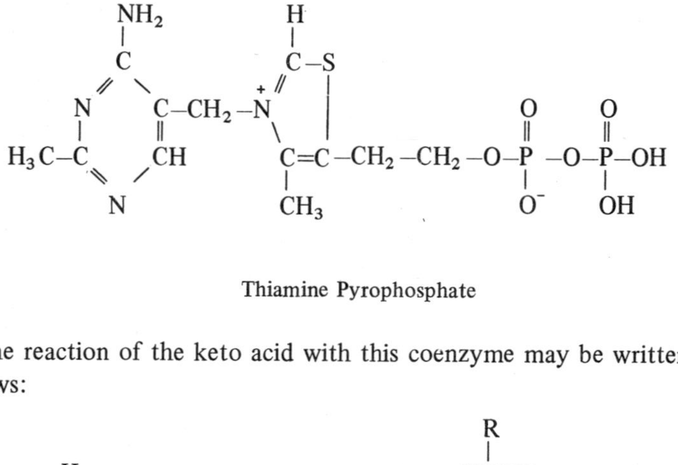

Thiamine Pyrophosphate

The reaction of the keto acid with this coenzyme may be written as follows:

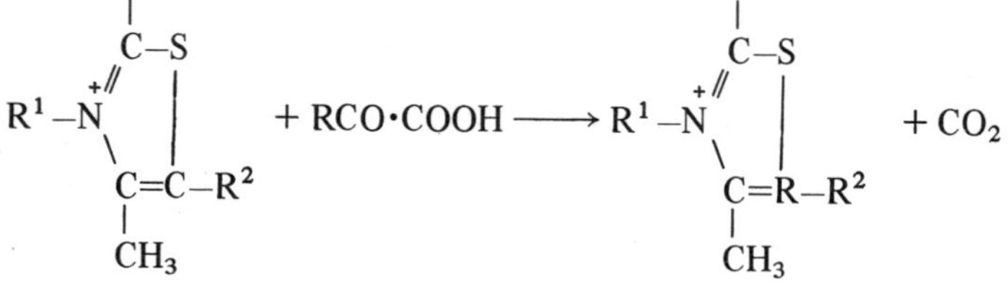

This reaction catalyzed by a decarboxylase results in the formation of a compound with thiamine pyrophosphate that serves as an "active aldehyde." An oxidation and a transfer step then occur in which lipoic acid plays the role of an electron acceptor. The exact details of this step are not known, but for simplicity in understanding it in the terms used here, it can be visualized as follows:

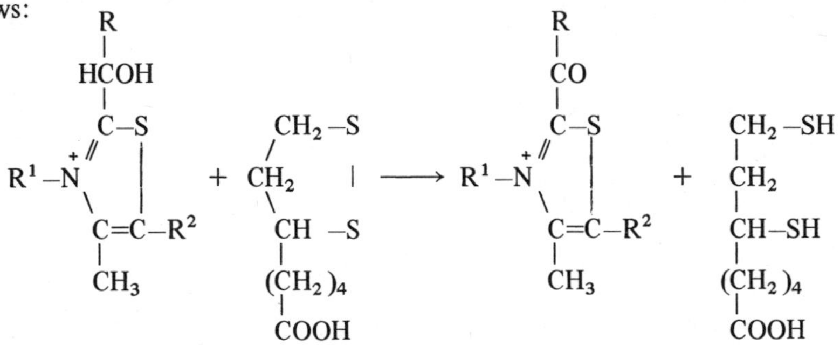

In this representation two hydrogen ions and two electrons are pictured as being removed from the –HCOH– group of the thiamine pyrophosphate complex and accepted by lipoic acid. A transfer reaction involving the ·CO–R group can then be visualized as occurring between the products to produce

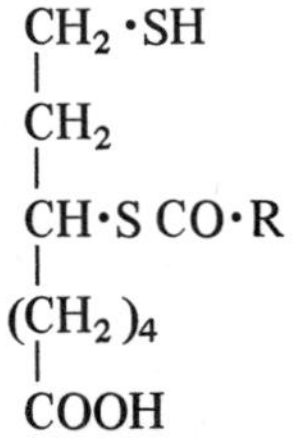

and regeneration of thiamine pyrophosphate. It must be emphasized that the dehydrogenation and transfer reactions may occur concomitantly so that the intermediate products depicted are never formed. The lipoic acid is not free as written. Its carboxyl group forms an amide linkage, $-\overset{\overset{O}{\|}}{C}-\overset{\overset{H}{|}}{N}-CH_2-R$, with the ε-amino group of a lysyl residue in the enzyme complex.

An exchange reaction next occurs in which the R·CO group attached to the lipoyl moiety of the enzyme complex is transferred to coenzyme A:

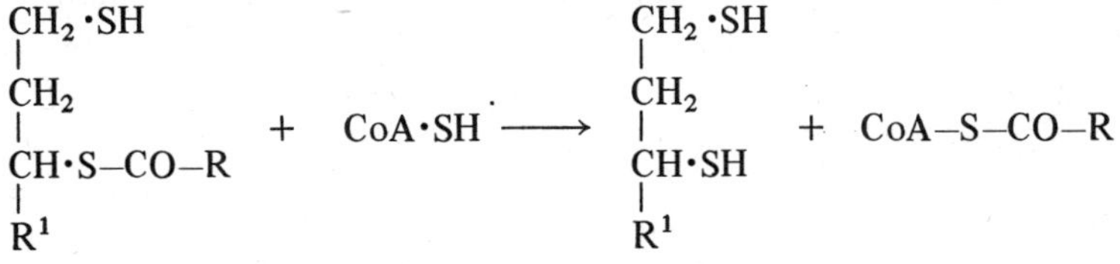

This results in the formation of succinyl CoA in the case of the dehydrogenation of α-ketoglutaric acid in the citric acid cycle (step 6, Fig. 2-1). When pyruvic acid derived from glucose (Chapter 3) undergoes oxidation, the product is acetyl CoA. The coenzyme A which participates in this transfer reaction is not permanently attached to the enzyme complex, and the succinyl or acetyl derivates of CoA which are formed are readily dissociated from the complex so that they may serve as substrates in other enzymatic reactions.

The next step in the chain of reactions is the dehydrogenation of the reduced lipoyl moiety:

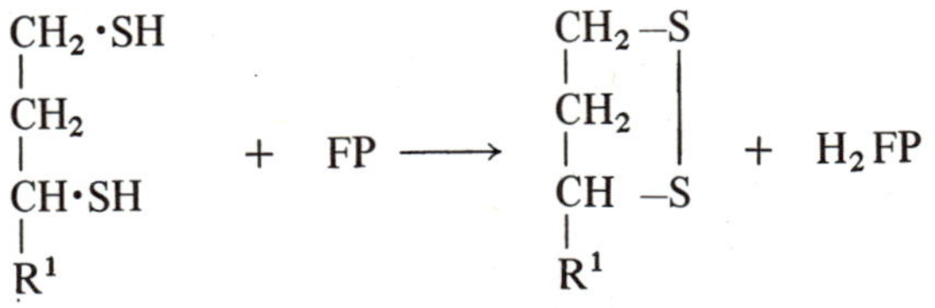

in which two electrons and two hydrogen ions are accepted by a flavoprotein and the disulfide form of lipoic acid is regenerated. Flavin adenine dinucleotide is the coenzyme of this flavoprotein enzyme.

The overall reaction to this point may be written as follows:

$$\mathrm{R{\cdot}CO{\cdot}COOH + FP + CoA{\cdot}SH \longrightarrow H_2FP + CoA{\cdot}S{\cdot}CO{\cdot}R + CO_2}$$

The final step in the process now remains the oxidation of H_2FP. In the test tube the following reaction can occur:

$$\mathrm{DPN^+ + H_2FP \longrightarrow DPNH + FP + H^+}$$

and it is presumed that this reaction also occurs in the living cell. It should be noted, however, that the reduction of DPN^+ by a reduced flavoprotein is an unusual reaction, and in most cases the reverse reaction is the one most likely to occur. This is because the oxidation–reduction potential of most flavoprotein systems is positive to that of the DPN^+–DPNH system.

In the case of the oxidation and decarboxylation of an α-keto acid, the end product, succinyl CoA or acetyl CoA, contains a thioester linkage. This linkage is written here as $-\overset{\overset{\displaystyle O}{/\!/}}{C}\sim S{\cdot}CoA$, since it is of the high-energy variety. In the case of acetyl CoA, the energy in this linkage was used to effect the condensation reaction with oxalacetic acid to start the citric acid cycle. In the case of succinyl CoA the energy is harvested for general purposes by a reaction in which inorganic phosphate (P_i) participates to form succinyl phospate,

$$\mathrm{COOH{\cdot}CH_2{\cdot}CH_2{\cdot}\overset{\overset{\displaystyle O}{/\!/}}{C}\sim OPO_3H^-}$$

The high-energy phosphate in this compound is then transferred to guanosine diphosphate (GDP) to produce guanosine triphosphate (GTP). This is the only reaction (7) in which GTP is known to be produced as a result of a biological oxidation. Reactions of this type are spoken of as high-energy phosphate production at the substrate level. They will be encountered again when we consider the conversion of glucose to acetyl CoA, but in these cases ATP will be generated rather than GTP. It should be noted, however, that ATP can be produced from GTP by the following reaction:

$$\mathrm{GTP + ADP \rightleftharpoons ATP + GDP}$$

The succinic acid formed from α-ketoglutaric acid undergoes a dehydrogenation reaction (8) with a flavoprotein containing FAD acting as the primary electron acceptor. The fumaric acid so produced then reacts with H_2O in a "make-ready" reaction (9) to yield malic acid. Another dehydrogenation (step 10) now occurs at the –CHOH– group of malic acid with DPN^+ as the electron acceptor. As a result, oxalacetic acid is regenerated for reaction with another molecule of acetyl CoA, and the cycle repeats itself. Thus the series of reactions from succinic acid to oxalacetic acid are a duplication of the type that was encountered in the oxidation of fatty acids to acetyl CoA.

Oxalacetic acid is essential to the functioning of the citric acid cycle. If any component of the citric acid cycle is used for another purpose, then regeneration of the oxalacetic acid by means of the cycle ceases. Components of the cycle which may be drained off include citrate, ketoglutaric acid, succinyl CoA, malic acid, and oxalacetic acid. Citrate, as will be seen later, plays a role in fatty acid synthesis in certain tissues. Succinyl CoA can be employed in porphyrin synthesis. Malate may be used to transfer reducing equivalents across the mitochondrial membrane. Ketoglutaric and oxalacetic acids are capable of being converted to the amino acids, glutamic and aspartic. The chief mechanism employed to replenish oxalacetate by those tissues in which there occurs such a diversion of citric acid cycle components involves fixation of CO_2 by pyruvate:

$$\mathrm{CH_3CO{\cdot}COO^- + CO_2 + ATP \longrightarrow HOOC{\cdot}CH_2{\cdot}CO{\cdot}COO^- + ADP + P_i}$$

This energy-requiring reaction is catalyzed by the enzyme pyruvate carboxylase which contains the vitamin biotin and requires Mg^{++}. This reaction will be seen subsequently to play an important role in fatty acid synthesis and gluconeogenesis. It should be noted that muscle contains very little pyruvate carboxylase, but this tissue is also deficient in those pathways that consume citric acid cycle components except perhaps for the conversion of the keto acids to amino acids. It is therefore important to recognize that conversion of the keto acids to glutamate and aspartate is a reversible reaction. Thus glutamate and aspartate can, when needed, serve to supply oxalacetate for the functioning of the citric acid cycle in tissues.

The reactions outlined for the conversion of a fatty acid completely to CO_2 may be summarized as follows, using palmitate as an example:

$$C_{16}H_{32}O_2 \longrightarrow 8CH_3CO{\cdot}S{\cdot}CoA + 7H_2FP + 7DPNH \qquad 7H^+$$
$$8CH_3CO{\cdot}S{\cdot}CoA \longrightarrow 16CO_2 + 8H_2FP + 24DPNH + 8 \sim P + 24H^+$$

$$\text{Sum:} \quad C_{16}H_{32}O_2 \longrightarrow 16CO_2 + 15H_2FP + 31DPNH + 8 \sim P + 31H^+$$

The reduced flavoproteins and DPNH do not accumulate in the cell, but hand the electrons they have accepted over to the electron transmitter system for final acceptance by oxygen. This process will be dealt with in a later chapter. The hydrogen ions formed in these initial reactions also do not accumulate. They may be temporarily buffered by the components of the cell, but they all eventually end up as H_2O, and no overall change in the pH of the cell occurs as a result of the oxidation process. A calculation of the total high-energy phosphate formed by the oxidation of one molecule of palmitate may be made from this equation. As a rule of thumb, each DPNH molecule may be considered to generate 3 $\sim$P and each H_2FP to generate 2 $\sim$P when they hand their electrons over to the electron transmitter system for reduction of O_2. On this basis the 31 molecules of DPNH will yield 93 $\sim$P and the 15 molecules of H_2FP will yield 30 $\sim$P, or a total of 123 $\sim$P by oxidative phosphorylation. To this must be added the 8 $\sim$P formed at the substrate level in the citric acid cycle, yielding a total of 131 $\sim$P. Since 2 $\sim$P are consumed in the process whereby the fatty acid is activated to form acyl CoA, the net yield is 129 $\sim$P per molecule of palmitate.

CHAPTER 3

Glucose Oxidation to Acetyl CoA

Entrance of glucose into the major metabolic pathways of the cell requires that it first be converted into glucose-6-phosphate. This reaction requires the expenditure of energy in the form of ATP:

$$\text{Glucose} + \text{ATP} \xrightarrow{Mg^{++}} \text{Glucose-6-}PO_4 + \text{ADP}$$

The enzyme catalyzing the reaction is called a hexokinase (glucokinase), and it may be inhibited by relatively low concentrations of glucose-6-PO_4. As is true of all kinase reactions, it involves the participation of Mg^{++}, which appears to form a chelate with the ATP. Four types of hexokinase are known. These are referred to as hexokinases I, II, III, and IV, depending on the rate of their electrophoretic movement at pH 8.4, type I moving the slowest. Liver contains all four types, while other tissues contain only two or three types. Of major interest is the fact that the different types of hexokinase display markedly different K_m values for glucose. Type III with a K_m value of 5.0×10^{-6} M has the highest affinity for glucose, while type IV with a value of 1.6×10^{-2} M has the lowest. Thus type III hexokinase is half-saturated with glucose at a concentration of approximately 0.09 mg per 100 ml, while for type IV this value is 288 mg per 100 ml, well above the normal serum glucose level of 90 mg per 100 ml. The formation of type IV hexokinase appears to be dependent on the presence of insulin.

The formation of the negatively charged glucose-6-phosphate molecule serves to trap glucose within the cell. Three pathways are then available for its metabolism:

$$\begin{array}{ccc} \text{Glycogen} & & \\ \downarrow\uparrow & & \\ \text{Glucose-6-phosphate} & \rightleftarrows & \text{Pyruvate} \\ \downarrow & & \\ \text{Pentose cycle} & & \end{array}$$

The pathway to glycogen will be treated later when energy storage reactions are discussed. The one leading into the pentose cycle is important for the generation of TPNH and will be considered in connection with fatty acid synthesis. The pathway leading to the formation of pyruvate is the one important in energy production from glucose and will concern us at this time.

The initial steps in the conversion of glucose-6-phosphate to pyruvate are shown in Fig. 3-1. According to the terminology we have adopted, all these may be called "make-ready" reactions. They entail first the conversion of glucose-6-phosphate to fructose-6-phosphate by the action of the enzyme phosphohexose isomerase. Then follows the expenditure of energy in the form of ATP for the formation of fructose-1,6-diphosphate. The activity of phosphofructokinase which catalyzes this reaction is subject to control by a number of compounds. Substances such as inorganic phosphate, ADP, and AMP, which tend to increase in amount when energy consumption is high, activate the enzyme and thus permit the more rapid flow of glucose into this energy-producing channel. Cyclic AMP, which, as will be seen later, is the major mobilizer of energy stores, also activates this enzyme. On the other hand, ATP and citrate serve as signals to the enzyme to diminish its activity. Both hexoseisomerase and phosphofructokinase require magnesium ions to function.

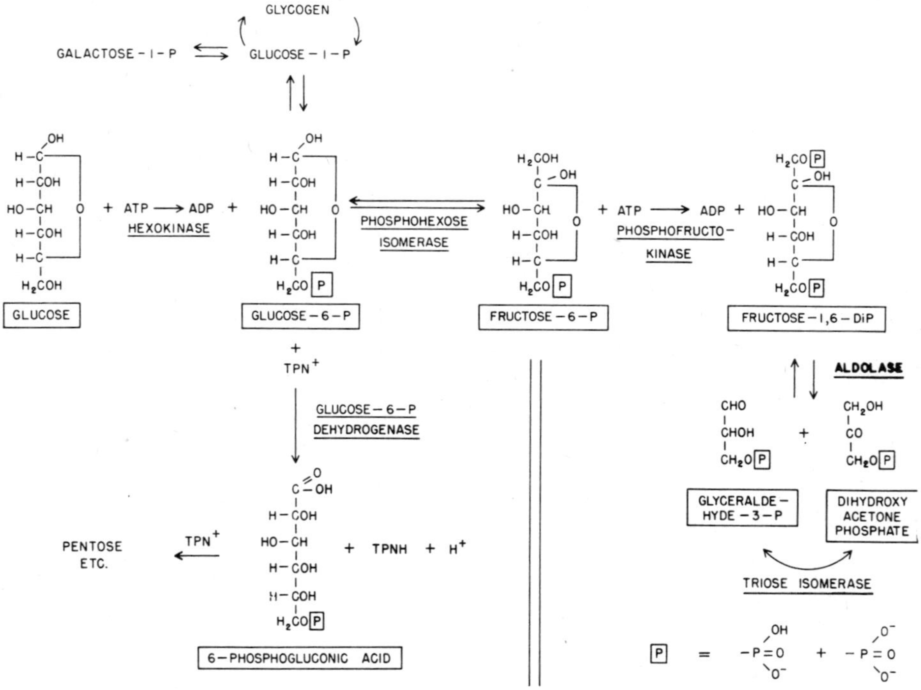

FIG. 3-1. The conversion of glucose to triose phosphates.

The final step in this series of "make-ready" reactions is the cleavage of fructose-1,6-diphosphate into the triose phosphates, dihydroxyacetone phosphate and glyceraldehyde-3-phosphate. The aldolase which catalyzes this reversible reaction possesses a lysyl residue which forms a labile Schiff base with the portion of fructose-1,6-diphosphate that yields dihydroxyacetone phosphate:

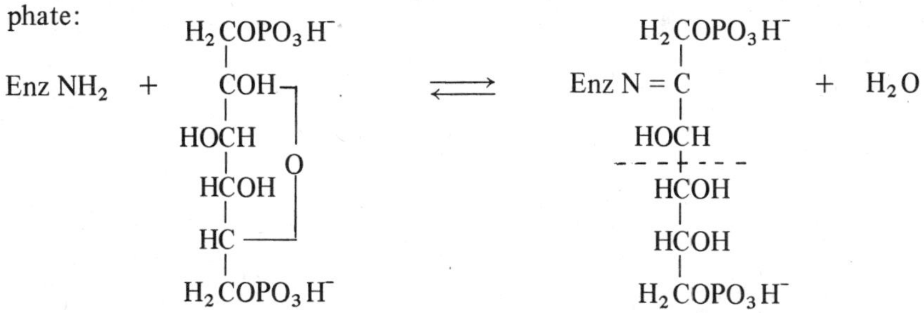

The triose phosphates formed by this cleavage are convertible one into the other through the catalytic action of the enzyme triose isomerase. At this point our primary interest will be in glyceraldehyde phosphate, since it is the compound that can undergo a dehydrogenation reaction. It should be pointed out, however, that dihydroxyacetone phosphate may either give rise to glyceraldehyde phosphate or be reduced to glycerophosphate as follows:

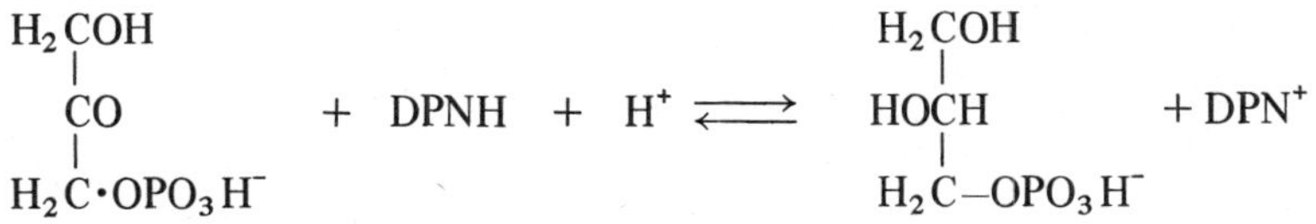

This reversible reaction is catalyzed by the enzyme glycerophosphate dehydrogenase. The production of glycerophosphate by this reaction is important in the synthesis of triglycerides and possibly in the transport of reducing equivalents from the cytoplasm to the mitochondria. The reverse reaction plays a role in the metabolism of glycerophosphate derived from glycerol. These aspects will be dealt with in more detail later.

Glyceraldehyde phosphate dehydrogenase, which catalyzes the next step (reactions 1 and 2, Fig. 3-2), is an interesting enzyme and one that has been subjected to much study. The enzyme as isolated from rabbit muscle has a molecular weight of 120,000 and is composed of three subunits of 40,000 molecular weight. There are four SH groups per subunit, and three of these react with SH reagents, such as iodoacetate, to yield an inactive enzyme. Treatment of the enzyme with ^{14}C-labeled iodoacetate and subsequent degradation of the enzyme has shown that the reactive SH group belongs to a cysteinyl residue. Unlike most dehydrogenases which have DPN^+ as a coenzyme, glyceraldehyde phosphate dehydrogenase tends to bind DPN^+ firmly. Some evidence has been presented to indicate that the enzyme SH groups may be involved in this binding. The exact mechanism of the dehydrogenation reaction remains somewhat in dispute. In presenting it in Fig. 3-2, we have represented the substrate as reacting with one of the SH groups of the enzyme, RSH, to form a thiohemiacetal linkage. A reaction of the substrate with the SH groups of the enzyme would be supported by the fact that these groups do not react with iodoacetate in the presence of the substrate. In any case, right or wrong, such a formulation achieves the production of the –CHOH– group that we have associated with dehydrogenation reactions involving DPN^+. It has been suggested that the DPNH produced in this reaction may not dissociate from the dehydrogenase, but passes two electrons and H^+ on to a molecule of DPN^+ in the medium.

The dehydrogenation reaction (2) produces a high-energy bond, $-\overset{O}{\overset{/\!/}{C}}\sim S-R$, of the same type as that encountered in the citric acid cycle with the formation of succinyl CoA. As in that reaction, this grouping is converted into a high-energy phosphate bond by reaction (3) with inorganic phosphate. As a result, 1,3-diphosphoglyceric acid is produced, and this in turn reacts (4) with ADP to yield ATP and 3-phosphoglyceric acid. This production of ATP at the substrate level is catalyzed by the action of the enzyme 3-phosphoglyceric acid kinase.

There then follows a set of reactions which are singular in that the production of high-energy phosphate occurs without a concomitant oxidation reaction. The first reaction involves the conversion of 3-phosphoglyceric acid to 2-phosphoglyceric acid catalyzed by phosphoglyceromutase and Mg^{++}. The removal of water through the action of an enolase then produces phosphoenolpyruvic acid. This compound possesses a high-energy phosphate bond with a value of about 14,000 calories and thus greater than the 8000 to 9000 calories assigned to those present in ATP. Consequently in the presence of pyruvate kinase, Mg^{++}, and ADP, the conversion of phosphoenolpyruvate to pyruvate and ATP occurs readily (6). The reverse reaction is not favored because of this difference in energy values, and this fact becomes important, as we shall see when the process of glucogenesis from pyruvate is discussed. Pyruvate kinase is dependent on K^+ ions as well as on Mg^{++} ions for activity.

The pyruvate formed is oxidized to acetyl CoA by pyruvate dehydrogenase (7). This process resembles that outlined in Chapter 2 for α-ketoglutarate in the citric acid cycle and involves the participation of thiamine, lipoic acid, flavin adenine dinucleotide, and CoA. The details will not be repeated here. The complex, pyruvate dehydrogenase, carrying out this reaction is reported to have a molecular weight in the vicinity of 5×10^6. The overall reaction is

$$\text{Pyruvate} + \text{CoA} + \text{FP} \longrightarrow \text{Acetyl CoA} + H_2\text{FP} + CO_2$$

The acetyl CoA produced enters the citric acid cycle, and thus the pathways of fatty acid and glucose oxidation become the same from this point on. In the test tube the reduced flavoprotein produced will react with DPN^+ to produce DPNH, and it is presumed that a similar reaction occurs within the cell.

Pyruvate dehydrogenase exists in two forms. It can react with ATP in the presence of a specific kinase to form a phosphorylated form of the enzyme which is inactive. It is reactivated by a phosphatase which splits off the phosphate. This type of control of enzyme activity will be presented in more detail when energy storage and mobilization reactions are discussed. As in these cases, there is evidence that hormones play a role in the initiation of the conversion of pyruvate dehydrogenase from one form to the other. The possible physiological significance of this control will be dealt with when glucogenesis is considered.

Under anaerobic conditions pyruvic acid formed from glucose may be converted into lactic acid. This reduction of pyruvate can be accomplished by the DPNH formed during the oxidation of glyceraldehyde phosphate (reaction 2) and in the presence of lactic acid dehydrogenase. Such a process is called anaerobic glycolysis; it occurs in muscle during periods of high muscular activity. The cell thus obtains energy from glucose without the participation of oxygen. The overall reaction may be written as follows:

$$\text{Glucose} + 2\,\text{ATP} \longrightarrow 2\,\text{Lactic acid} + 4\,\text{ATP}$$

In yeast, pyruvate is decarboxylated to yield acetaldehyde which is then reduced to ethyl alcohol (reactions 8 and 10). The cell achieves a net gain of 2 ~P per molecule of glucose converted to lactate or alcohol. This may be contrasted with the yield of high-energy phosphate obtained when glucose is converted to CO_2 and water.

$$\begin{array}{rlrl}
\text{Glucose} & \longrightarrow 2\,\text{Pyruvate} & + & 2\,\text{DPNH} \qquad\qquad + 2\sim\text{P} \\
2\,\text{Pyruvate} & \longrightarrow 2\,\text{Acetyl CoA} + 2\,\text{CO}_2 & + & 2\,\text{DPNH} \\
2\,\text{Acetyl CoA} & \longrightarrow \qquad\qquad 4\,\text{CO}_2 & + & 6\,\text{DPNH} + 2\,\text{H}_2\text{FP} + 2\sim\text{P} \\
\hline
\text{Sum: Glucose} & \longrightarrow \qquad\qquad 6\,\text{CO}_2 & + & 10\,\text{DPNH} + 2\text{H}_2\text{FP} + 4\sim\text{P}
\end{array}$$

If we assume (see Chapter 4) that each DPNH yields 3 ~P and that H_2FP yields 2 ~P when their electrons are fed into the electron transmitter system to react with oxygen, then the total yield of ~P is 38 for each molecule of glucose completely oxidized. Thus anaerobic glycolysis yields only about 5% of the energy from glucose that is obtained by its complete combustion. It may also be noted that only 4 of the 38 ~P, or roughly 10% of the total energy production during complete glucose oxidation, is obtained by high-energy phosphate production at the substrate level.

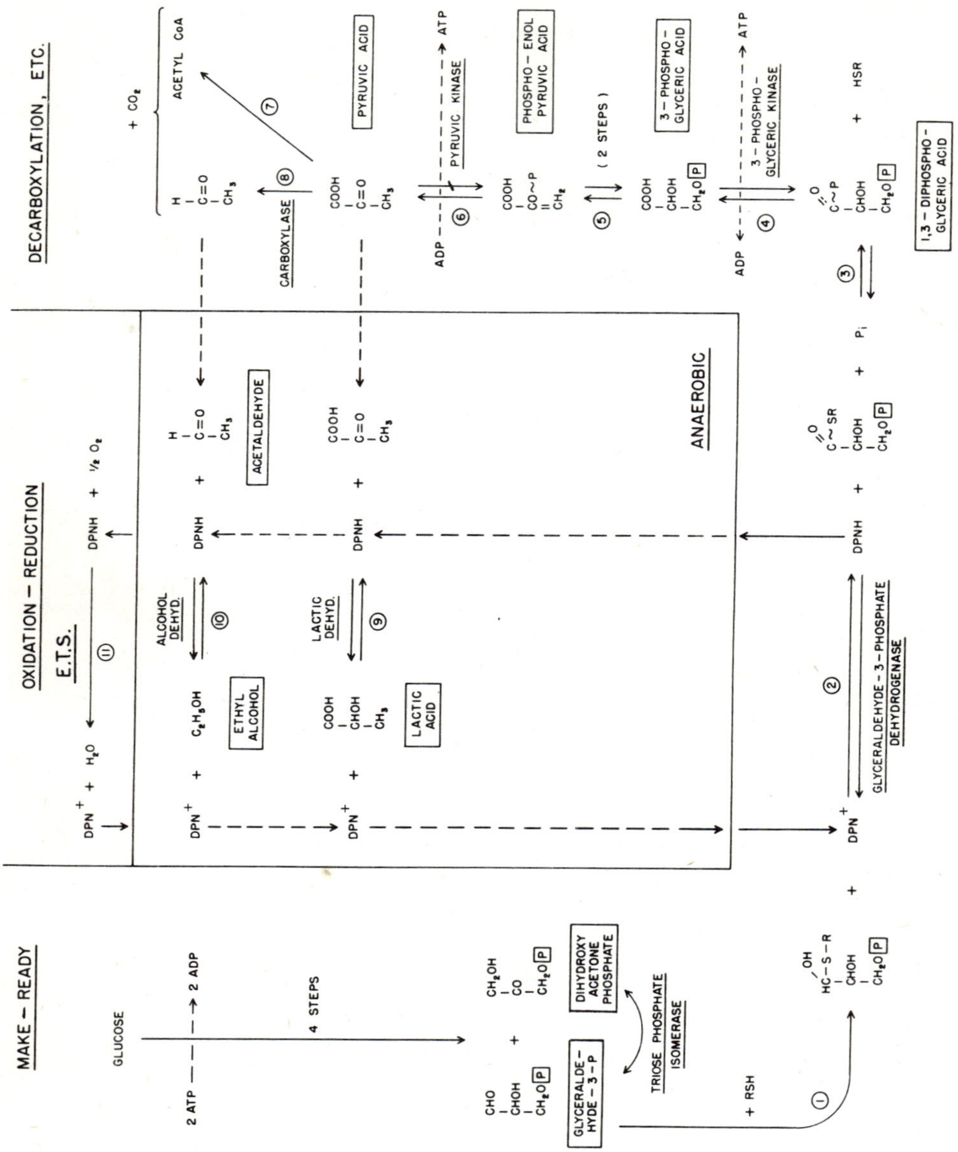

FIG. 3-2. The oxidation of glucose to pyruvate and the anaerobic formation of lactate or alcohol.

CHAPTER 4

Electron Transmitter System

In dealing with the series of reactions in which glucose and fatty acids are converted to CO_2, we have followed the transfer of electrons from the substrates to the primary electron acceptors, DPN^+ and flavoproteins. We now turn to the process whereby the reduced forms of these acceptors hand over their electrons to the electron transmitter system for conveyance to oxygen and the formation of water. This system is embedded in the inner mitochondrial membrane and, in contrast to the dehydrogenases, is insoluble in nature. For convenience we shall refer to this system as E.T.S. Its known components are the iron porphyrin compounds, cytochromes b, c, and c_1, and cytochrome oxidase; a number of flavoproteins; and a quinone, called coenzyme Q or ubiquinone. The mitochondrial membrane containing these components is also known to contain about 40% phospholipids, iron not chelated in the cytochromes and so-called structural proteins, as well as components concerned in oxidative phosphorylation. The fact that the E.T.S. is insoluble and is composed of so many compounds which act together in an integrated manner has presented difficulties in the unraveling both of its composition and of the sequence of reactions that take place. As a consequence some uncertainties still exist about the exact nature and role of certain components.

Each component of the E.T.S. is capable of existing in an oxidized and a reduced form and hence constitutes an oxidation–reduction system with an oxidation–reduction potential. In Fig. 4-1 the various components, named for the oxidant, are arranged along a potential scale which reflects their relative relationships with regard to the oxidation–reduction systems which each component forms. Theoretically the oxidant of any system should be capable of accepting electrons from the reductant of any system with a lower oxidation–reduction potential. Hence on this basis one can predict that, if any exchange of electrons between the components listed in Fig. 4-1 is to occur, it must be in the order right to left. However, it must be emphasized that, though the potential difference is favorable, an interaction of components may not readily

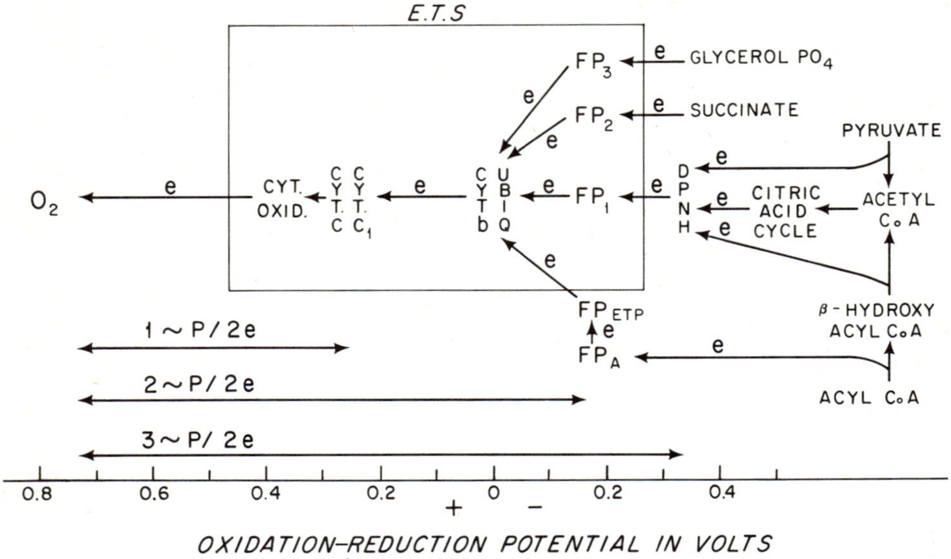

FIG. 4-1. The pathway of electron flow (*e*) within the mitochondria. The compounds enclosed by the solid line form part of the mitochondrial membrane. Of these the components cytochrome oxidase (cyt. oxid.); cytochromes c, c_1, and b(cyt. c, c_1, b); ubiquinone (ubiq.); and the flavoprotein (FP_1) form the electron transmitter system (E.T.S.). The flavoproteins (FP_2) and (FP_3) are, respectively succinate dehydrogenase and glycerolphosphate dehydrogenase. Each of these compounds is the oxidant of an oxidation–reduction system and is so positioned on the diagram that it lies approximately above that point on the voltage scale which corresponds to the normal oxidation–reduction potential at pH 7.0 of that system. The other compounds depicted are present in the mitochondrial matrix. Placement of these compounds on the diagram has significance with regard to oxidation–reduction potentials only in the case of the systems formed by DPNH (diphosphopyridine nucleotide); FP_{ETP} (electron transport protein) and FP_A (acyl CoA dehydrogenase).

occur. For example, from the respective potentials of their systems, the oxidant of cytochrome c should be able to accept electrons from DPNH; however, this does not occur. In general there is an amazing specificity of interaction between the components of the E.T.S., so that an ordered sequence of electron exchange occurs. The E.T.S. is like a series of locks in a canal where each lock must be taken in turn. Electron flow in the E.T.S. thus proceeds stepwise from one component to another. The order of the sequence in which interaction occurs is a matter which can only be determined experimentally, but by and large it is found to be consistent with the order given by the oxidation–reduction potentials.

In addition to the specificity of interaction that exists between components of the E.T.S., it must be remembered that each component of the E.T.S. is fixed in the mitochondrial membrane. Thus the spatial relationship in this membrane of one component to another should also be a determining factor in the order of their interactions. Thus it is as if nature had made "assurance double sure" that electron flow through the E.T.S. follow a fixed pathway and that no short circuit of electrons occur. One reason that can be given for the use of such an ordered sequence by the cell is that it permits the stepwise harvesting of energy. As indicated in Fig. 4-1, the flow of electrons into and through the E.T.S. yields three high-energy phosphates, each of which is formed at different locations (called sites I, II, and III) along the chain of oxidation–reduction reactions. Without such an ordered sequence of electron flow, the formation of multiple and identical packets of energy in the form required by the cell might not be achieved.

Entry of electrons into the E.T.S. from the various possible sources would appear always to be through a flavoprotein. In the case of DPNH this seems clearly to be the case, and it has even been proposed that two flavoproteins working in series may be involved. The flavoproteins which act as primary acceptors in the oxidation of acyl CoA derived from fatty acids do not appear to hand over their electrons directly to the E.T.S. Instead, a so-called electron-transferring flavoprotein appears to act as an intermediate in this case.

Embedded in the mitochondrial membrane which contains the E.T.S., one may find two flavoproteins which accept electrons directly from substrate molecules. One of these, present in all tissues, is succinate dehydrogenase. As we have seen, this enzyme is an integral part of the citric acid cycle. Why this and only this dehydrogenase of the citric acid cycle is made an integral part of the mitochondrial membrane containing the E.T.S. system is a mystery. In a few tissues one may also find a flavoprotein, glycerol phosphate dehydrogenase, as a component of E.T.S. preparations. This enzyme is present in large amounts in certain tissues and almost absent in others. The activity of this enzyme in tissues such as liver is greatly influenced by the thyroid hormone. It is high in the hyperthyroid state and low in hypothyroid conditions. It has been proposed that this enzyme is involved in an extra-intramitochondrial electron shuttle system which will be discussed in more detail later.

The flow of electrons along the E.T.S. to oxygen from any one of the numerous reduced flavoproteins that have been mentioned cannot be pinpointed exactly. Either cytochrome b or ubiquinone would appear to be involved. It is possible that one of these components acts in one case and the other in another. However, it is of interest to note that ubiquinone is much more plentiful in the E.T.S. system than cytochrome b. It is generally agreed that the four cytochrome components of the E.T.S. are present in nearly equal molar amounts. On a molar basis ubiquinone has been shown to be present in about five times the amount of cytochrome c. Since the cytochromes can accept only one electron per mole and ubiquinone two, this means that the ubiquinone in the E.T.S. would have about ten times the electron-carrying capacity of cytochrome b. Such a relationship would favor the notion that ubiquinone plays a primary role in the reaction with the various reduced flavoproteins. This is because it would then be possible for each of the reduced flavoproteins to react with different ubiquinone molecules. Since the flavoproteins are relatively large molecules and are embedded presumably in a fixed position in the mitochondrial membrane, it is difficult to envision one cytochrome b molecule oriented spatially so as to achieve the same role.

Recent work indicates that two forms or species of cytochrome b exist. These have been designated as cytochrome b_T and cytochrome b_K or cytochrome b_{565} and cytochrome b_{562}. The subscripts T and K employed by one group of workers stand for "transduction" and "Keilin". In the other nomenclature the subscripts stand for the wavelength of maximum absorption by the reduced forms as measured by the investigators who employ this terminology. Evidence has been presented that each of these two forms participates in a separate pathway of electron flow but in the same relative position to the other components of the E.T.S. as is shown for cytochrome b in Fig. 4-1. However, passage of electrons over cytochrome b_K does not result in the production of high-energy phosphate at site II as is the case when cytochrome b_T is involved. In this connection the effect of ATP on the oxidation–reduction potentials of the systems formed by the two types of cytochrome b is of interest. In the absence of ATP the potential of both systems is in the vicinity of zero volt as shown for cytochrome b in Fig. 4-1. In the presence of ATP the potential of the system formed by cytochrome b_T, but not cytochrome b_K, shifts to a value of +0.25 volt and thus into the region of the potentials displayed by the cytochrome c systems. The participation of cytochrome b_T in the formation of high-energy phosphate (transduction) is thus suggested (see Chapter 5).

Relative to the other components of the E.T.S., ubiquinone is a small molecule with the following basic structure:

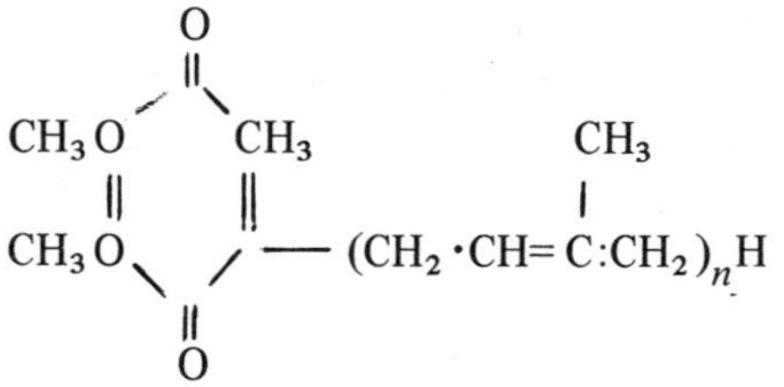

In mammalian tissues *n* is usually 9 to 10, but in unicellular organisms *n* may have lower values. It is insoluble in water, but can be dissolved in fat solvents. This suggests that its presence in the mitochondrial membrane is dependent on the high content of phospholipids found there.

If ubiquinone serves as a gathering point for the electrons donated by all the reduced flavoproteins formed in the E.T.S., then cytochrome b is the next component that must be considered for participation in the flow of electrons. It is important to note that, up to this point, all the participating components of the E.T.S. have been two-electron carriers. The remainder of the components of the E.T.S. are cytochromes which are one-electron carriers ($Fe^{3+} + e \longrightarrow Fe^{++}$). Hence the reaction of reduced ubiquinone with cytochrome b (or cytochrome c or c_1) requires either that two cytochrome molecules react with one molecule of reduced ubiquinone to form ubiquinone, or that reduced ubiquinone yield one electron at a time for passage along the remainder of E.T.S. This latter event is not impossible, since it is known that certain quinones can readily form free radicals called semiquinones by the acceptance of a single electron.

Cytochromes c and c_1 are both candidates for the next step in the sequence of electron flow to oxygen. The role of cytochrome c_1 is more controversial than that of cytochrome c. There seems little doubt that cytochrome c is an essential component of the E.T.S. It is possible to extract cytochrome c from preparations of the E.T.S. by relatively gentle methods. Such extracted preparations still contain cytochrome c_1 and the other components of the E.T.S., but they have lost their capacity to transmit electrons from DPNH or succinate to O_2. The addition of cytochrome c restores this ability. Reduction of cytochrome c_1 by either succinate or DPNH can be shown to occur in the preparations freed of cytochrome c. It would thus appear that, if cytochrome c_1 plays a role in the transfer of electrons, it must be between cytochrome c and ubiquinone or cytochrome b.

Cytochrome c is the most studied component of the E.T.S. This is in large part due to the fact that it was the easiest and the first component to be obtained in pure form. It has a molecular weight of 13,000 and an isoelectric point of 10.6, which gives it a positive charge at physiological pH values. This fact may help to account for its relatively loose attachment to the mitochondrial membrane with its large number of negatively charged phospholipid molecules. The amino acid sequence of cytochrome c prepared from a large variety of living forms has been made in studies directed toward the unraveling of genetic changes that occur along the phylogenetic tree. The ability of cytochrome c to act as an oxidation–reduction system is due to its iron protoporphyrin nucleus.

This is the same grouping found in cytochromes b and c_1 and in hemoglobin. However, unlike hemoglobin, cytochromes b, c, and c_1 do not combine with oxygen, and their ferrous forms are not readily converted to the ferric state by oxygen.

The final component in the chain of compounds transferring electrons to oxygen is cytochrome oxidase. This component is perhaps the most complex of the E.T.S. Some workers believe that cytochrome oxidase is composed of two components called cytochrome a and cytochrome a_3, which contain iron chelated to a porphyrin different in structure from that in cytochrome b or c.

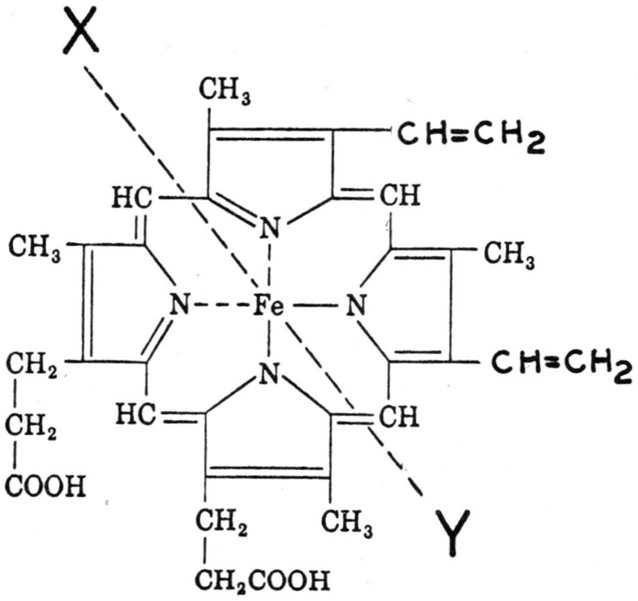

Iron Protoporphyrin
X, Y = points of ligand formation with the protein portion of a cytochrome.

In addition, most investigators now consider copper to be an integral part of cytochrome oxidase. In considering the action of cytochrome oxidase it is important to bear in mind that the reduction of O_2 to $2H_2O$ which it achieves requires four electrons, and also that each molecule of reduced cytochrome c, which presumably provides these electrons, can furnish but one electron. Lemberg has proposed that each unit of cytochrome oxidase is composed of two iron porphyrin groups and two copper molecules, so that in abbreviated form its reduced structure can be written as $Fe_a^{++} Cu^+ + Fe_b^{++} Cu^+$, where Fe_a^{++} and Fe_b^{++} represent iron porphyrin groups. One of the iron porphyrin groups is envisaged to combine with oxygen to yield $Fe_a^{++} O_2 Cu^+ + Fe_b^{++} + Cu^+$ in much the same manner as it does to the iron prophyrin group in hemoglobin. Also as in hemoglobin, CO can displace O_2 from its attachment and thus account for the ability of CO to act as a respiratory poison. Cytochrome oxidase has long been known to react with CO. Otto Warburg made use of the ability of light to dissociate CO from cytochrome oxidase to establish by spectroscopic methods that it contained an iron porphyrin compound. It was the shift in part of the absorption spectra of reduced cytochrome a in the presence of CO which led Keilin to suggest that cytochrome oxidase was composed of two components, which he named cytochrome a and cytochrome a_3.

Unlike hemoglobin, which holds O_2 at arm's length and prevents its ferrous iron from conversion to ferric, cytochrome oxidase rapidly undergoes an oxidation with the reduction of oxygen to water. Lemberg represents this reaction as follows:

$$\underline{Fe_a^{++}O_2Cu^+ + Fe_b^{++}Cu^+} + 4H^+ \longrightarrow \underline{Fe_a^{4+}Cu^+ + Fe_b^{3+}Cu^{++}} + 2H_2O$$

The oxidized form of cytochrome oxidase is then envisaged as being reduced in two steps, each involving the acceptance of two electrons. The first step is the reduction of the $\overline{Fe_b^{3+} + Cu^{++}}$ portion of the molecule to form the intermediate $\underline{Fe_a^{4+}Cu^+ + \overline{Fe_b^{++}\ Cu^+}}$. There then follows the acceptance of two more electrons to regenerate the fully reduced cytochrome oxidase, $\underline{Fe_a^{++}\ Cu^+ + Fe_b^{++}\ Cu^+}$. In each step presumably two reduced cytochrome c molecules would serve as electron donors.

As was mentioned earlier, the mitochondrial membrane which contains the E.T.S. is relatively rich in iron which is not chelated to the cytochromes. This so-called nonheme iron is attached to relatively small protein molecules which are rich in cysteinyl residues (SH groups). Reduction of this iron is accompanied by a characteristic electron spin resonance signal. The participation of nonheme iron as an electron carrier, especially in association with certain flavoproteins, is strongly suggested by recent studies. Reduction of the iron in all the cytochromes is accompanied by the appearance of sharp and intense absorption bands in both the visible and ultraviolet regions of the spectrum. For example, in the visible region the reduced cytochromes show bands centered at the following wavelengths: cytochrome oxidase ($a + a_3$), 605 mμ; cytochrome b, 563 mμ; cytochrome c, 550 mμ. It is perhaps worth noting here that the original assignment of the letters a, b, and c by Keilin to the cytochromes was based on the relative positions of their absorption bands and was made some years before the order of their participation in the E.T.S., namely a, c, and b, had been established.

The fact that all the components of the E.T.S. exhibit characteristic absorption bands in either their reduced or oxidized forms has been of enormous value in elucidating the role of any one component in the chain of electron flow. Coupled with the use of spectrophotometric methods, especially as perfected by Chance and his co-workers, much fundamental information has been yielded by the employment of substances that inhibit electron flow at specific points in the chain. Among the inhibitors that have been employed, cyanide was perhaps one of the earliest to be used. Cyanide reacts with cytochrome oxidase, presumably with its ferric iron, to block oxygen uptake of the cell. It can be shown that in the presence of cyanide all components of the E.T.S. except cytochrome oxidase will remain reduced and cannot be converted to their oxidized forms by oxygen. Other useful inhibitors are the antibiotic antimycin A, which blocks the passage of electrons from cytochrome b to c; and rotenone, an insecticide, and sodium amytal, a barbiturate drug, both of which inhibit electron flow between DPNH and cytochrome b. The employment of electron acceptors other than oxygen has also been a useful tool. In the absence of oxygen or in the presence of inhibitors, electron flow can occur over various portions of the E.T.S. from substrates to a variety of inorganic or organic substances which form oxidation–reduction systems. For example, methylene blue can be reduced to its colorless form in the absence of oxygen, or the dye can serve to link the E.T.S. to oxygen when cyanide has been used to poison cytochrome oxidase.

CHAPTER 5

Oxidative Phosphorylation

The flow of electrons from the primary electron acceptors through the electron transmitter system is, under normal circumstances, accompanied by the production of high-energy phosphate by the conversion of ADP and P_i to ATP. This process is called oxidative phosphorylation. Indeed there is usually such a tight coupling between the process of electron flow and oxidative phosphorylation that the former does not occur without the latter. Thus consumption of oxygen is regulated by energy demands and specifically by the production of ATP from ADP. The term respiratory control is applied to this regulatory process.

The process of oxidative phosphorylation may be studied in fragments of the mitochondrial membrane which contain the E.T.S. Such preparations, however, do not generally show the tight coupling between electron flow and phosphorylation that can be observed in intact mitochondria. In Fig. 5-1 a schematized diagram is presented which illustrates respiratory control. Here oxygen consumption is represented as a fall in oxygen tension as measured by an oxygen electrode. If carefully prepared mitochondria are added to a suitably buffered solution containing inorganic phosphate, a small oxygen consumption may be observed. Since the mitochondria contain the enzymes of the citric acid cycle (see Chapter 6), it is possible to use substrates of this cycle as electron donors – for example, succinate or α-ketoglutarate. The addition of such a substrate results in little or no change in the O_2 tension. If, however, a small amount of ADP is added, the substrate undergoes dehydrogenation and an increase in O_2 consumption occurs. When all the ADP has been converted to ATP, electron flow ceases and the rate of oxygen consumption drops again to the rate observed before ADP addition. The measurement of the amount of oxygen consumed during this period along with the amount of ADP converted to ATP provides a value which is expressed as $\sim$P/O, the amount of high-energy phosphate formed per atom of oxygen . Since two electrons must flow through the E.T.S. for each oxygen atom reduced to water, the ratio $\sim$P/2*e* may be used.

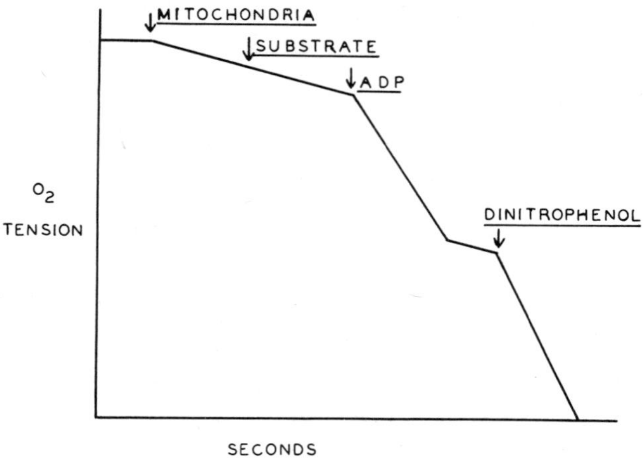

FIG. 5-1. A schematic representation of respiratory control and the uncoupling of oxidative phosphorylation.

Also illustrated in Fig. 5-1 is the process in which release of respiratory control or uncoupling of electron flow and oxidative phosphorylation can be produced by the addition of a suitable compound. Used here as an illustration is dinitrophenol, the first compound known to produce this effect. A sizable list of compounds which can act as uncouplers is now known. Among these may be mentioned thyroxine and fatty acids, compounds of natural occurrence. The addition of an uncoupler permits the flow of electrons from substrate to oxygen to occur until one or the other is consumed. Since energy storage in the form of ATP does not occur, the energy generated in the process of electron flow is dissipated as heat. It has been proposed that such a generation of heat occurs in brown adipose tissue by the uncoupling action of fatty acids. This tissue is extremely rich in mitochondria and is plentiful mainly in hibernators. The addition of catecholamines to this tissue *in vitro* produces a release of fatty acids and an enormous burst in oxygen consumption. A similar event is considered to occur in the animal during arousal from hibernation and to aid in bringing its body temperature more rapidly to normal.

The maximum value for the ∿P/O ratio is generally considered to be 3.0 when the electron donor is DPNH and 2.0 when it is a reduced flavoprotein. Such values may be determined as described above. For example, the use of α-ketoglutarate as a substrate in an experiment of the type depicted in Fig. 5-1 yields ∿P/O values approaching 4 when the inhibitor malonate is added to block the further oxidation of succinate in the citric acid cycle. Since the oxidation of α-ketoglutarate involves DPN^+ and one high-energy phosphate is produced at the substrate level in its conversion to succinate, it is argued that three ∿P must be produced by oxidative phosphorylation. DPNH cannot be used directly as an electron donor in such experiments, since the mitochondrial membrane appears to be impermeable to this compound (see Chapter 6). Similar experiments using succinate as substrate, where a flavoprotein is the primary acceptor, yield values approaching 2 for the ∿P/O ratio. When reduced cytochrome c is added as an electron donor, values approaching one for ∿P/O are obtained. The conclusion is therefore drawn from experiments of this type that three sites of phosphorylation exist. The formation of one high-energy phosphate bond thus is considered to occur between DPNH and flavoprotein or possibly cytochrome b or ubiquinone, another between these compounds and cytochrome c or c_1 and a third between cytochrome c and oxygen. Whether the last site is between cytochrome c and cytochrome oxidase, cytochrome oxidase and oxygen, or a combination of all three is not resolved (see Fig. 4-1).

It is possible to make some approximate calculations of energy yields in the process of oxidative phosphorylation from the oxidation–reduction potentials of the components of the E.T.S. For example, the oxidation–reduction potential of the DPN–DPNH system is –0.318 volt. If the value of +0.810 volt is used for the potential of the standard oxygen electrode, then a voltage drop of 1.128 volts may be said to occur as electrons flow from DPNH to O_2. The free energy, ΔG°, of this reaction may then be calculated from the equation $\Delta G^\circ = -nFE$. Substituting a value of 1.128 volts for E, 23,068 for F, the Faraday, and a value of 2 for n, the number of electrons flowing, a value of −52,000 calories is obtained for ΔG°. Now if three high-energy phosphate bonds with a ΔG° of 8000 calories are formed during the passage of two electrons from DPNH to O_2, (∿P/O = ∿P/2e = 3), then 24,000 calories are produced as energy useful to the organism. On this basis the efficiency of the overall process of oxidative phosphorylation is therefore somewhat less than 50%.

The equation $\Delta G^\circ = -nFE$ may also be used to calculate the minimum voltage difference that must occur to produce one high-energy phosphate bond for the passage of two electrons. With the values given above, this yields a figure of −0.173 volt. The voltage difference between the standard potential of the DPN^+–DPNH system and that of any of the systems represented by flavoproteins, cytochrome b, or ubiquinone can be estimated to be of the order of 0.3 ± 0.05 volt; between the latter systems and the cytochrome c system, 0.25 ± 0.05 volt; and approximately 0.55 volt from cytochrome c to O_2. On this basis the last step in oxidative phosphorylation is the least efficient. These calculations are based on the flow of two electrons. A voltage drop of 0.346 volt would be required to produce one high-energy phosphate if only one electron is involved. This fact becomes important to consider in relation to the central role played by cytochrome c which is a one-electron system.

The total energy output of the average human at rest may be calculated on the basis of electron flow. The oxygen consumption for such an individual may be taken as 240 ml/min, or 6.5×10^{21} molecules. Since four electrons are required to reduce one molecule of O_2 to H_2O, this means an electron flow of 26×10^{21} electrons/min. Now 3.76×10^{20} electrons/min. equals 1 ampere; hence the total electron flow is 69 amperes. If we assume for the sake of simplicity that all electrons traverse the total pathway from DPNH to O_2, the voltage drop may be taken as 1.13 volts. The total energy release is then equivalent to 1.13×69, or 80 watts. An electric light bulb consuming 80 watts can be used for 24 hours at a cost between 5 and 10 cents. One pound of carbohydrate would supply the basic energy needs of a human for 24 hours at a comparable cost.

The exact process whereby high-energy phosphate is produced as electrons flow into and through the E.T.S. is not known. One of the commonly accepted premises as to the mechanism involved is represented by the following equation:

$$\text{Reductant A} + \text{X} + \text{oxidant B} \longrightarrow \text{Oxidant A}{\sim}\text{X} + \text{reductant B}$$

Here A and B represent unspecified members of the E.T.S. which interact with each other with the exchange of electrons. In the process some substance X, which is not known, is depicted as forming a high-energy derivative with the electron donor A as electrons pass to B. It is then postulated that one of the following reactions occurs:

$$\text{Oxidant A}{\sim}\text{X} + \text{P}_\text{i} \longrightarrow \text{Oxidant A}{\sim}\text{P}$$
$$\longrightarrow \text{Oxidant A} + \text{X}{\sim}\text{P}$$

in which inorganic phosphate reacts to form a high-energy phosphate compound with either oxidant A or substance X. The last step is then either of the following:

$$\text{Oxidant A}{\sim}\text{P} + \text{ADP} \longrightarrow \text{Oxidant A} + \text{ATP}$$
$$\text{X}{\sim}\text{P} + \text{ADP} \longrightarrow \text{X} + \text{ATP}$$

Further participation of components A and B in electron transmission is considered impossible as long as either oxidant ~P or X~P exists. Respiratory control is thus exerted, and initiation of electron flow will occur only when ADP is present to accept high-energy phosphate and to regenerate oxidant A or X. One may picture uncoupling agents as acting in some manner so as to cause the hydrolysis of oxidant A~X, oxidant A~P, X~P, or ATP with the generation of heat and uncontrolled electron flow.

This scheme for the process of oxidative phosphorylation finds its counterpart in the process for high-energy phosphate production at the substrate level.

For example, if reductant A is replaced by glyceraldehyde phosphate, X by HS-Enz, the dehydrogenase, and oxidant B by DPN^+, the pattern is that in which oxidant$\sim$X would become

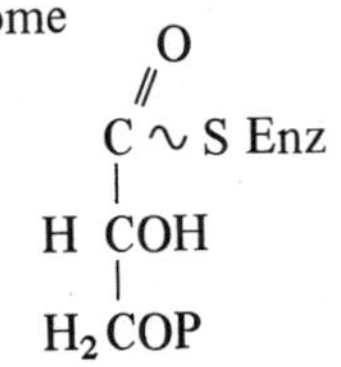

In this case the next step is the formation of A$\sim$P rather than X$\sim$P. If the process of oxidative phosphorylation is patterned on that for production of high-energy phosphate at the substrate level, then one difference exists. Agents that uncouple oxidative phosphorylation do not interfere with phosphorylation at the substrate level.

A number of modifications of this basic form of what is called the chemical-coupling hypothesis of oxidative phosphorylation have been proposed. One, for example, introduces one or more enzymes (E), or protein carriers of high energy between the oxidant $\sim$X and the end product ATP.

$$\text{Oxidant A}\sim\text{X} + \text{E} \rightleftharpoons \text{Oxidant A} + \text{E}\sim\text{X}$$

$$\text{E}\sim\text{X} + \text{P}_i \rightleftharpoons \text{E}\sim\text{P} + \text{X}$$

$$\text{E}\sim\text{P} + \text{ADP} \rightleftharpoons \text{E} + \text{ATP}$$

The main support for this modification comes from the identification of so-called coupling factors in mitochondrial membrane fragments which display oxidative phosphorylation. Removal of these factors from the fragments results in uncoupling—that is, the loss of ATP production and the free flow of electrons from substrate to oxygen. On addition of these coupling factors, phosphorylation can be restored. At least five coupling factors, all proteins, have been described. One of these factors, F_1, has been obtained in Racker's laboratory as a homogeneous protein and has a molecular weight around 320,000. It displays the property of catalyzing the hydrolysis of ATP to ADP and P_i and is considered to be involved in the final step of oxidative phosphorylation:

$$\text{E}\sim\text{P} + \text{ADP} \rightleftharpoons \text{E} + \text{ATP}$$

Whether more than one coupling factor is involved in the overall process and whether each of the three sites of oxidative phosphorylation requires different coupling factors remain unresolved.

The scheme presented for oxidative phosphorylation calls for the formation of a high-energy derivative of a component of the E.T.S. In mammalian systems the identification of such a derivative and the unknown compound X has remained elusive. Recently reports have appeared that changes in the oxidation–reduction potentials of cytochrome b and cytochrome a_3 can occur that would suggest the existence of energy-transducing forms of these electron carriers. The inability to identify a chemical-coupling compound has led to the introduction of other theories. One, called the chemiosmotic hypothesis, proposes that the flow of electrons along the E.T.S. generates a gradient of H^+ across the mitochondrial membrane and that this in turn supplies the energy for ATP formation. Support for this hypothesis sponsored by Mitchell is found in the fact that respiration-dependent accumulation of various cations by mitochondria can be observed. The reader is referred to the article by Greville given in the references for a more extended discussion of this hypothesis.

As in studies of the E.T.S., inhibitors have been useful tools in probing the mechanism of oxidative phosphorylation. As was mentioned earlier in this chapter, various agents are known which act as uncouplers. Most of these prevent ATP formation but permit the free flow of electrons through the E.T.S. A special class of such uncouplers are compounds like the antibiotics gramicidin and valinomycin, which exert uncoupling effects upon mitochondria only in the presence of K^+ or some other monovalent cation. The energy released in electron flow in such cases is apparently used to pump the cation inwardly across the mitochondrial membrane. At the same time an equivalent amount of H^+ ions is released from the interior of the mitochondria into the medium. An interesting inhibitor of oxidative phosphorylation is the antibiotic oligomycin. Blockage of high-energy phosphate production by this compound does not result in unrestrained electron flow. However, if dinitrophenol is also added, then true uncoupling results. It has been possible to isolate a factor called F_0 which must be present in order for oligomycin to exert its action on the phosphorylation processes as observed in mitochondrial fragments. The information supplied by the use of compounds of this sort can and has been used in support of both the chemical-coupling and chemiosmotic theories. Obviously, much remains to be learned about the process of oxidative phosphorylation. As one worker in this field has facetiously remarked about the process, "Anyone who is not thoroughly confused, just does not understand the situation."

CHAPTER 6

Cellular Localization of Reactions

In presenting the metabolic pathways in the previous chapters, no attention has been paid to the localization within the cell of these reactions. Before proceeding further, it is worth while to consider this aspect, since it emphasizes problems of transport of reactants across cell membranes. The pinpointing of the location of a given enzyme or process within the cell can entail some uncertainties, even though the process of disrupting a cell and separating its contents has been highly refined in recent years. In this regard the following remarks made in 1960 by Dame Honor B. Fell in her presidential address before the International Society for Cell Biology are still worthy of quotation.

"Cell structure was more and more disregarded and the reign of the homogenizer began. Cells were savagely disembowelled and people hurried to investigate the biochemical properties of the isolated organelles. Surprisingly (to a morphologist like myself) a lot of interesting and valuable information emerged from this holocaust. But naturally the results were one sided, and the risk of physiological artifact was considerable. Who could tell how far the functional activities of a mitochondrion sitting cozily in the living cytoplasm would resemble those of a mitochondrion torn from its natural home and exiled in a completely abnormal environment" (H.B. Fell, Fashions in Cell Biology, *Science* **132**, 1625-1627 (1960). Copyright 1960 by the American Association for the Advancement of Science).

The localization of an enzymatic process in the cell may be studied by the technique of homogenization and differential centrifugation. In this process tissue cells—for example, liver—are disrupted by suspending pieces of the tissue in an isotonic sucrose or KCl solution and subjecting them to shearing forces produced by the rotation of a pestle in a relatively close-fitting glass tube. The various membrane-bound particles of the cell called organelles are thereby released from the cell, some in a more damaged state than others. The suspension obtained is called an homogenate. After filtering off any gross particles such as blood vessels, connective tissue, etc., the homogenate is subjected to a stepwise series of centrifugations at increasing rotor speeds so as to produce increasing gravitational forces (*g*). For example, the nuclei are found in the sediment after spinning for 10 minutes at 600 *g*. The supernatant from this step will yield mitochondria if it is spun at 1500 *g* for about 5 minutes. The homogenate is further cleared of all organelles by a third centrifugation at 100,000 *g* for 60 minutes, yielding a fraction called the microsomes and containing ribosomes and fragments of the endoplasmic reticulum. The final supernatant has been called the cytosol. It should be noted that the terms microsomes and cytosol are not used by the histologists in describing the contents of the intact cell. Enzymes found in the cytosol are presumed to be located in the cytoplasmic matrix of the intact cell and not to be components of any of the organelles. However, the possibility that enzymes are released into the cytosol from organelles damaged during the homogenization process must always be taken into consideration. Fortunately the techniques of the histochemist have done much to substantiate the conclusions drawn from the homogenization procedures.

Most of the reactions that have been considered in the previous chapters are reported to occur within the mitochondria. The main exception are those reactions in which glucose is converted to pyruvate. The overall reactions listed according to their cellular localization are as follows:

Cytosol or Extramitochondrial

$$\text{Glucose} \longrightarrow \text{DPNH} + \text{Pyruvate} + \sim\text{P}$$

$$\text{Fatty acid} + \text{CoA} + \text{ATP} \longrightarrow \text{Acyl CoA} + \text{AMP} + \text{Pyro P}$$

Intramitochondrial

$$\text{Pyruvate} \longrightarrow \text{Acetyl CoA} + CO_2 + \text{DPNH}$$

$$\text{Acyl CoA} \longrightarrow \text{Acetyl CoA} + \text{DPNH} + H_2\text{FP}$$

$$\text{Acetyl CoA} \longrightarrow CO_2 + \text{DPNH} + H_2\text{FP} + \sim\text{P}$$

$$H_2\text{FP} + \text{DPNH} + O_2 \longrightarrow H_2O + \text{FP} + \text{DPN} + \sim\text{P}$$

A consideration of the localization of these reactions is of interest from the standpoint of logistics. First it means that all production of CO_2 and utilization of O_2 by these reactions occur within the mitochondria. Both O_2 and CO_2 must traverse the mitochondrial membranes as well as the cell membrane on their way from or to the bloodstream. Both these gases are more soluble in lipids than in water. Since all membranes are rich in phospholipids, the passage of O_2 and CO_2 through these membranes should present no problems. It would seem reasonable to conclude that the interior of the mitochondria has the highest CO_2 tension and perhaps a more acidic pH than the rest of the cell. Also an important intramitochondrial reaction which should be favored by a high CO_2 tension is the following:

$$\text{Pyruvate} + CO_2 + \text{ATP} \longrightarrow \text{Oxalacetate} + \text{ADP}$$

The importance of the production of oxalacetate by this reaction for the formation of citrate in the citric acid cycle may be recalled. Its significance in fatty acid synthesis and glucogenesis will be apparent in later chapters.

Second, it can be seen that the major production of DPNH and H_2 flavoprotein occurs within the mitochondria. Both these substances feed electrons into the E.T.S. which is embedded in the inner mitochondrial membrane. This inner membrane, especially in tissues in which energy production is high, forms numerous cristae which interlace the matrix of the mitochondria. Thus the E.T.S. is conveniently located to accept electrons from these primary electron acceptors and rapid and unhindered reaction should be possible. In the case of the DPNH produced outside the mitochondria, as in the conversion of glucose to pyruvate, the situation is different. Penetration of the mitochondria by external DPNH does not appear to occur, so that direct reaction with the E.T.S. is not possible. If the conversion of glucose to pyruvate is to occur, then some means must be available for the DPNH produced in this reaction to unload its electrons. This may be accomplished by a so-called shuttle mechanism in which the DPNH reacts to reduce a metabolite which then carries the electrons into the mitochondria. Several ways in which this may be achieved have been suggested. First, the possibility exists that pyruvate is reduced to lactate and that lactate rather than pyruvate penetrates the mitochondria. It has been reported that lactate dehydrogenase is present in the mitochondria so that both pyruvate and DPNH could be regenerated inside the mitochondria. Second, a shuttle system can be formed with glycerophosphate as the electron carrier. In this system the DPNH formed by the oxidation of glyceraldehyde phosphate is coupled with the reduction of the other triose phosphate formed from glucose, namely dihydroxyacetone phosphate. The glycerol phosphate formed then enters the mitochondria where it is oxidized back to dihydroxyacetone phosphate by means of the flavoprotein glycerol phosphate dehydrogenase, which, as has been mentioned, is embedded with the E.T.S. system in the mitochondrial membrane of certain tissues. It is of interest that this specific dehydrogenase is plentiful in the mitochondria of insect wing muscles which are, strangely enough, deficient in lactate dehydrogenase. Since this shuttle system involves the generation of a reduced flavoprotein for DPNH, it is possible, though not proved, that a sacrifice of one high-energy phosphate would be entailed in its use. Finally, a third shuttle system involving β-hydroxybutyrate as electron carrier has been proposed. Though rigid proof for the functioning of any of these shuttle systems is lacking, the possibility that they exist or that different tissues employ one or more of them cannot be ignored. When we come to consider the synthesis of fatty acid from glucose, we shall encounter another possible manner by which electrons may be removed from DPNH generated outside the mitochondria.

The third point that needs consideration is that the majority of the high-energy phosphate production occurs within the mitochondria and largely in the form of ATP. On the other hand, many of the energy-consuming reactions occur outside the mitochondria. These include, for example, the synthesis of fatty acids, proteins, nucleic acid, and other cellular components. The use of ATP for muscular contraction, nerve conductance, maintenance of ionic gradients, and many other cellular functions also must take place outside the mitochondria. The cell must therefore possess means for permitting the energy generated within the mitochondria to be made available to all parts of the cell. Evidence is accumulating that this may be accomplished by the outward flow of ATP and inward flow of ADP across the mitochondrial membranes.

Finally, in considering logistics of the cell in relation to energy production, there needs to be mentioned the role of carnitine, $(CH_3)_3 -N^{+} \cdot CH_2 \cdot CHOH \cdot CH_2 \cdot COOH$, in the transport of fatty acids into the mitochondria. As we have seen earlier, the first step in the preparation of fatty acids for oxidation is the formation of their CoA derivatives. This process occurs outside the mitochondria. The membranes of the mitochondria appear to be impermeable to acyl CoA, but permeable to acyl carnitine. An enzyme is found in both the cytosol and the mitochondria which catalyzes the reaction:

$$\text{Acyl CoA} + \text{carnitine} \rightleftharpoons \text{Acyl carnitine} + \text{CoA}$$

in which the –CHOH– group of carnitine forms an ester linkage with the fatty acid. Entrance of the acyl group into the mitochondria is thus effected. Entrance of fatty acid into the cell's cytoplasm from the bloodstream does not seem to present any special problem.

CHAPTER 7

Energy Storage

Since many animals, including man, eat at periodic intervals, it is necessary for them to store away materials that can be mobilized for their energy needs during inanition. Glycogen and fat are the two main materials used for this purpose, and both are derived chiefly from ingested carbohydrates. In this chapter we shall consider the processes whereby glycogen and fats are formed from glucose.

Fat is the preferred form for energy storage by animals where the need for mobility is best served by a substance with the highest energy yield per gram. Glycogen yields only 43% of the energy provided by the same amount of fat. Moreover, glycogen storage is accompanied by the storage of water, which is not the case for fat. This fact is reflected in the relative amounts of glycogen and fat found in a normal individual. For example, in a man weighing 70 kg, 10% of the body weight or 7 kg may be stored as fat. For man at rest this amount of fat would supply his basic energy needs for over a month. Glycogen is mainly stored in the liver and muscle. In liver, glycogen can amount to 4% of its wet weight. In a 70-kg man with a 1.75-kg liver this would provide 70 g of glycogen. The glycogen content of muscle is of the order of 1% of its wet weight. If the total muscle mass is taken as 28 kg, then some 280 g of glycogen is stored in this tissue. The total glycogen in these two organs thus amounts to 350 g, an amount that would provide a human with his total basic energy needs for about 1 day.

Since the formation of glycogen is a somewhat simpler process than fatty acid synthesis, it will be presented first. The first step in the conversion of glucose to glycogen is the same as in the preparation of glucose for oxidation – namely, the formation of glucose-6-phosphate. Then follows the conversion of glucose-6-phosphate to glucose-1-phosphate. This reaction is reversible and may be written as occurring in the following manner:

$$\text{Glucose-6-PO}_4 + \text{Enz O PO}_4 \rightleftharpoons \text{Enz OH} + \text{Glucose-1,6-PO}_4 \rightleftharpoons \text{Glucose-1-PO}_4 + \text{Enz O PO}_4$$

The enzyme involved (Enz OH), is called phosphoglucomutase. It contains a seryl residue the hydroxyl group of which can react with glucose-1,6-diphosphate to form a phosphorylated enzyme (Enz O PO_4) and either glucose-1-PO_4 or glucose-6-PO_4. A small amount of glucose-1,6-diphosphate thus needs to be present to initiate the reaction if the enzyme is in the unphosphorylated state. Should production of glucose-1,6-diphosphate be required, the cell achieves this by the following reaction:

$$\text{Glucose-1-PO}_4 + \text{ATP} \longrightarrow \text{Glucose-1,6-PO}_4 + \text{ADP}$$

The enzyme concerned is called phosphoglucokinase and resembles phosphofructokinase in its action.

The second step in glycogen formation is the reaction of glucose-1-PO_4 with uridine triphosphate (UTP) to form uridine diphosphate glucose and pyrophosphate.

$$\text{Glucose-1-PO}_4 + \text{UTP} \longrightarrow \text{UDP glucose} + \text{PP}_i$$

The pyrophosphate formed is hydrolyzed to inorganic phosphate, and the formation of UDP glucose is thus favored. The last step is then the reaction of UDP glucose with glycogen, which may be represented as follows:

$$\text{UDP glucose} + \text{glycogen } (n \text{ glucose units}) \longrightarrow \text{Glycogen } (n{+}1 \text{ glucose units}) + \text{UDP}$$

This reaction is catalyzed by the enzyme UDP glucose–glycogen glucosyl transferase, more conveniently called glycogen synthetase. The UDP formed in this reaction can be reconverted to UTP by reaction with ATP:

$$\text{UDP} + \text{ATP} \longrightarrow \text{UTP} + \text{ADP}$$

The conversion of glucose to glycogen thus requires the expenditure of two high-energy phosphate bonds per glucose molecule, one for the reaction above and the other for the formation of glucose-6-PO_4.

The enzyme glycogen synthetase (GS) can exist in a phosphorylated and a nonphosphorylated form. The interconversion of the enzyme from one form to the other controls the rate of glycogen synthesis:

$$GS + ATP \longrightarrow GS{\cdot}PO_4 + ADP$$

Both forms of the enzyme are capable of glycogen synthesis. The phosphorylated form is the least active and moreover requires the presence of glucose-6-phosphate to be active. This dependence upon glucose-6-phosphate for activity has led to $GS{\cdot}PO_4$ being called the dependent or D form of the enzyme, while GS, which does not require glucose-6-phosphate for activity, is called the independent or I form. The rate of glycogen formation from glucose in muscle has been shown to increase as the ratio I/D increases.

The phosphorylation and hence the inactivation of glycogen synthetase are catalyzed by an enzyme called glycogen synthetase kinase. The activity of this enzyme is under hormone control. For example, in muscle insulin depresses its activity and epinephrine increases it. Thus the presence of insulin favors glycogen synthesis, and the release of epinephrine retards it. A more detailed explanation of the mode of action of these hormones and the role of cyclic AMP in the activation of kinases of this type will be presented in Chapter 8.

We shall turn now to the consideration of the process whereby glucose is converted to fatty acids. The initial step in this process is the conversion of glucose to acetyl CoA; it thus follows the same pathway that was outlined in our consideration of glucose oxidation. The acetyl CoA, however, instead of undergoing oxidation in the citric acid cycle, becomes reduced to fatty acids by the action of TPNH. Note that this conversion of acetyl CoA to fatty acids occurs in the cytoplasm, while acetyl CoA production is an intramitochondrial process. The translocation of the acetyl CoA across the mitochondrial membrane presents a special problem which will be dealt with later.

The conversion of acetyl CoA to fatty acids resembles basically, but in reverse, the process followed in the oxidation of fatty acids to acetyl CoA. A series of stepwise reactions couples together acetyl CoA units with two reductive steps after the addition of each two carbon units. The process repeats itself until an acyl chain unit of sixteen, eighteen, or more carbon atoms has been reached. The coupling and reductive reactions are carried out, however, with the acyl unit attached to a special protein rather than to coenzyme A as in the oxidative sequence. This protein, called acyl carrier protein (ACP), has a molecular weight of about 10,000, and is an analog of coenzyme A in that the protein replaces the 3′,5′-adenosine phosphate portion of coenzyme A. The 4′-phosphopantetheine portion of coenzyme A is linked to a seryl residue of ACP through a phosphodiester linkage. The first step in the series of reactions is an exchange of an acetyl group between the SH groups of coenzyme A and ACP catalyzed by the enzyme acetyl transacylase:

$$CH_3CO{\cdot}S\ CoA + HS{\cdot}ACP \longrightarrow CH_3CO{\cdot}S{\cdot}ACP + HS\ CoA$$

The next step in the series of reactions is the addition of another acetyl CoA unit to $CH_3CO{\cdot}S{\cdot}ACP$ to produce the four-carbon acetoacetyl derivative of the carrier protein, $HOOC{\cdot}CH_2CH_2CO{\cdot}S{\cdot}ACP$. To effect this condensation the cell employs a device which ensures that the reaction is exergonic, so that the synthetic process is strongly favored. This is accomplished by the following series of reactions. Malonyl CoA is formed from acetyl CoA by the fixation of CO_2:

$$CH_3{\cdot}CO{\cdot}S\,CoA + CO_2 + ATP \longrightarrow HOOC{\cdot}CH_2{\cdot}CO{\cdot}S\,CoA + ADP + P_i$$

This reaction is catalyzed by acetyl CoA carboxylase which contains covalently bound biotin. Fixation of CO_2 in this type of reaction can be represented as occurring in two steps:

$$HCO_3^- + ATP + \text{biotin enzyme} \rightleftharpoons {}^-OOC\text{–biotin enzyme} + ADP + P_i$$

$${}^-OOC\text{ biotin enzyme} + \text{acetyl CoA} \rightleftharpoons \text{Biotin enzyme} + \text{malonyl CoA}$$

The structure of the carboxybiotin enzyme intermediate may be formulated as

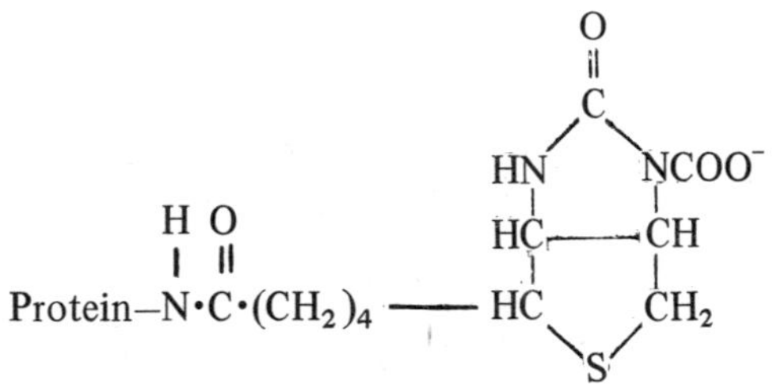

The malonyl CoA formed then reacts with another molecule of ACP in the same manner as for acetyl CoA:

$$HOOC{\cdot}CH_2{\cdot}CO{\cdot}S\,CoA + HS{\cdot}ACP \longrightarrow HOOC{\cdot}CH_2CO\text{–}S{\cdot}ACP + CoA\,SH$$

A condensation reaction now occurs involving a molecule of the acetyl and malonyl derivative of ACP to form acetoacetyl ACP. The release of the CO_2 previously fixed in malonyl CoA serves to drive this reaction to the right.

$$CH_3CO{\cdot}S{\cdot}ACP + HOOC{\cdot}CH_2CO{\cdot}S{\cdot}ACP \longrightarrow CH_3CO{\cdot}CH_2{\cdot}CO{\cdot}S{\cdot}ACP + HS{\cdot}ACP + CO_2$$

It should be noted that the process of fatty acid synthesis requires the presence of CO_2, though the CO_2 entering into the process does not appear in the final product. The sum of all these reactions may be written as

$$2 \text{ Acetyl CoA} + \text{ACP} + \text{ATP} \longrightarrow \text{Acetoacetyl ACP} + \text{ADP} + P_i + 2 \text{ CoA SH}$$

Reduction of acetoacetyl ACP to butyryl ACP then occurs by the following series of reactions:

$$CH_3 \cdot CO \cdot CH_2 \cdot CO \cdot S \cdot ACP + TPNH + H^+ \longrightarrow CH_3 \cdot CHOH \cdot CH_2 \cdot CO \cdot S \cdot ACP + TPN^+$$

$$CH_3 \cdot CHOH \cdot CH_2 \cdot CO \cdot S \cdot ACP \longrightarrow CH_3 \cdot CH = CH \cdot CO \cdot S \cdot ACP + H_2O$$

$$CH_3 \cdot CH = CH \cdot CO \cdot S \cdot ACP + TPNH + H^+ \longrightarrow CH_3 \cdot CH_2 \cdot CH_2 \cdot CO \cdot S \cdot ACP + TPN^+$$

This series of reactions, in the order given, is catalyzed by β-ketoacyl-ACP reductase, enoyl ACP hydrase, and crotonyl ACP reductase. These reactions are in essence the reverse of those that occur in the oxidation of the CoA derivatives of fatty acid. However, TPNH is used in both the reductive steps, in contrast to the use of DPN^+ and a flavoprotein in the oxidative process.

Elongation of the four-carbon acyl chain attached to ACP is now possible by the use of another molecule of malonyl ACP:

$$CH_3 \cdot CH_2 \cdot CH_2 \cdot CO \cdot S \cdot ACP + HOOC\ CH_2 \cdot CO \cdot S \cdot ACP \longrightarrow CH_3 \cdot CH_2 \cdot CH_2 \cdot CO \cdot CH_2 \cdot CO \cdot S \cdot ACP + HS \cdot ACP$$

The six-carbon β-ketoacyl ACP formed may then be reduced by TPNH in the same manner as that described for the four-carbon derivative. Further stepwise additions of malonyl ACP with reduction may then be repeated until the carbon chain length of the acyl group reaches fourteen, sixteen, or eighteen. In this process the methyl group of the initial acetyl ACP remains the terminus of the growing chain. The formation of palmitic acid containing sixteen carbon atoms from acetyl CoA by the process outlined may be summarized as follows:

$$8CH_3 \cdot CO \cdot S\ CoA + 14\ TPNH + 7\ ATP + HS \cdot ACP \longrightarrow CH_3 \cdot (CH_2)_{14} \cdot CO \cdot S \cdot ACP + 8\ HS\ CoA + 14\ TPN^+$$

The above equation tells only part of the story. We must now consider the additional requirements which are imposed upon fatty acid synthesis, first, by the fact that the conversion of acetyl CoA to fatty acids occurs in the cytosol, but acetyl CoA formation occurs in the mitochondria and second, by the need for TPNH production for the reductive steps in the process. As will be seen, these two facets of the synthetic process are not unrelated. The overall set of reactions that will be considered are presented in Fig. 7-1.

Formation of acetyl CoA occurs by the oxidation of pyruvate in the mitochondria. Outward diffusion of this acetyl CoA from the mitochondria for fatty acid synthesis does not apparently occur readily. The reverse process, it will be recalled, also does not take place, and passage of acetyl CoA into the mitochondria is effected by formation of acetyl carnitine. Although formation of acetyl carnitine within the mitochondria appears to occur, the cell apparently does not use this process to any great extent in the translocation of acetyl CoA for fatty acid synthesis. Instead, for a reason that will be discussed later, it utilizes an energy-consuming device for this purpose. The acetyl CoA is condensed with oxalacetate to form citrate by the same reaction as that described for the initiation of the citric acid cycle (reaction 3, Fig. 7-1). The citrate, however, instead of undergoing oxidation is released into the cytoplasmic matrix where it is converted back to acetyl CoA and oxalacetate in a reaction with requires ATP.

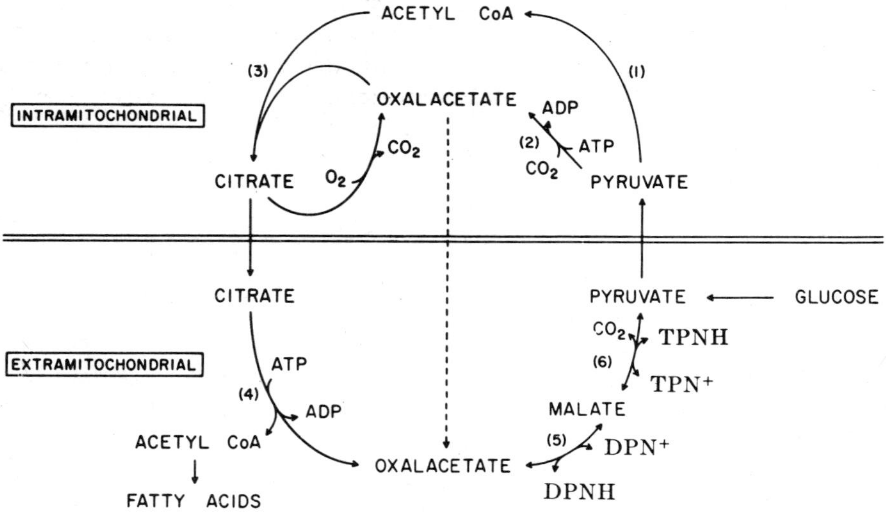

FIG. 7-1. Fatty acid formation from glucose. The reactions indicated are catalyzed by (1) pyruvate oxidase; (2) pyruvate carboxylase; (3) citrate condensing enzyme; (4) citrate cleavage enzyme; (5) malate dehydrogenase; and (6) malic enzyme. [Reproduced from M.S. Kornacker and E.G. Ball, *Proc. Nat. Acad. Sci. U.S.* 54, 899–904 (1965).]

$$\text{Citrate} + \text{ATP} + \text{CoA} \longrightarrow \text{Acetyl CoA} + \text{oxalacetate} + \text{ADP} + \text{P}_i$$

The requirement for ATP stems from the fact that regeneration of acetyl CoA requires the formation of the high-energy bond $CH_3C(=O){\sim}S\,CoA$ that was lost during the formation of citrate. This reaction is catalyzed by an enzyme called citrate lyase or citrate cleavage enzyme. This enzyme is found only in tissues such as liver and adipose which are capable of fatty acid synthesis.

The oxalacetate formed by the cleavage of citrate may be reduced to malate by DPNH (reaction 5, Fig. 7-1). This reaction is catalyzed by malate dehydrogenase, an enzyme that is present in both the mitochondria and cytosol of tissues in which fatty acid synthesis occurs. The extramitochondrial DPNH needed can be provided by the step in which glucose is converted to pyruvate, specifically by the oxidation of glyceraldehyde phosphate.

The malate formed may now react with TPN^+ to produce TPNH, pyruvate, and CO_2. The catalyst for this reaction is called malic enzyme or malate dehydrogenase–TPN^+ linked. The pyruvate formed mingles with that generated by the oxidation of glucose and can reenter the mitochondria. This process obviously creates a loss of oxalacetate needed for citrate formation, and its regeneration becomes necessary. This can be accomplished by the reaction catalyzed by pyruvate carboxylase (reaction 2, Fig. 7-1).

It must be emphasized that evidence for the participation of the series of reactions just outlined is at present largely of an indirect nature. For example, the three key enzymes involved – citrate lyase, malic enzyme, and pyruvate carboxylase – are most plentiful in tissues that carry out fatty acid synthesis. The activities of citrate cleavage and malic enzyme have been shown to rise markedly in tissues as the rate of lipogenesis is increased. In rats the rate of lipogenesis in liver and adipose tissue can be increased by starvation for 3 days followed by refeeding for 2 days. Concomitantly with the increased fat synthesis one finds that the activity of both these enzymes increases three- to fourfold over that seen prior to alteration of the dietary regime. Treatment of pigeons with prolactin for 3 to 5 days produces striking and parallel increases in the ability of their liver to synthesize fatty acids and the activities of malic enzyme and citrate lyase. Finally, it has been shown in the mammary gland that malic enzyme levels are elevated eightfold and citrate lyase is increased twelvefold with the onset of lactation.

Although the evidence for the operation of what may be called the malate cycle is not complete, certain aspects are worthy of note. First is the fact that each turn of this cycle requires that two high-energy phosphates be supplied, one for pyruvate conversion to oxalacetate and the other for citrate cleavage. At first glance this may seem wasteful. However, if the primary purpose of the cell is considered to be the conversion of glucose to fatty acids, the utilization of high-energy phosphate for these processes actually serves to facilitate conversion of glucose to fatty acids. The reasons for this may be seen by comparison of the reactions involved in the production of fatty acid from glucose which give rise to high-energy phosphate production with those that consume high-energy phosphate. The following reactions are those involved in high-energy phosphate production:

$$\text{Glucose} \longrightarrow 2 \text{ Pyruvate} + 2 \sim\text{P}$$

$$2 \text{ Pyruvate} \longrightarrow 2 \text{ Acetyl CoA} + 2 \text{ DPNH}$$

$$2 \text{ DPNH} + O_2 \longrightarrow 6 \sim\text{P} + 2 \text{ DPN}^+ + 2H_2O$$

The production of each acetyl CoA unit is thus accompanied by the production of four ∿P. On the other hand, energy-utilizing reactions are the following:

$$\text{Citrate} + \sim\text{P} \longrightarrow \text{Acetyl CoA} + \text{oxalacetate}$$

$$\text{Pyruvate} + CO_2 + \sim\text{P} \longrightarrow \text{Oxalacetate}$$

$$\text{Acetyl CoA} + CO_2 + \sim\text{P} \longrightarrow \text{Malonyl CoA}$$

These sum up to the utilization of three ∿P for each acetyl CoA going to fatty acid. (The value is slightly less than three, since the first acetyl CoA group incorporated into fatty acids is not carboxylated to malonyl CoA.) Thus in the conversion of glucose to fatty acids, four ∿P are formed for each acetyl CoA unit produced from glucose, but three of these are consumed in the processes by which acetyl CoA is converted to fatty acids. This utilization of high-energy phosphate thus becomes an important factor in keeping the overall process in motion. If these energy-consuming reactions did not occur, then the rate of fatty acid synthesis would be strictly limited by the rate at which the cell consumed high-energy phosphate for its basic energy requirements. Even so, some restriction must be imposed on fatty acid synthesis by the basic rate of energy phosphate utilization by the cell. This is because for each acetyl CoA unit four ∿P are produced and only three are utilized. If respiratory control is to be maintained, then some of the ∿P produced must be consumed by the cell to supply its basic energy requirements. That this is indeed the case can be shown in adipose tissue. The initiation of fatty acid synthesis in this tissue is accompanied by a sizable diminution in the activity of the citric acid cycle. Production of ∿P by operation of the citric acid cycle is thus curtailed, since the cell is being supplied with ∿P from this other source. At the same time more citrate becomes available for fatty acid synthesis.

It is possible that the importance of maintaining a balance between ATP production and utilization during fatty acid synthesis has not been adequately emphasized in discussions of the control of the rate of fatty acid synthesis. An important control point in this process could be the rate at which citrate is formed. The K_m of the enzyme citrate synthetase for acetyl CoA is increased as ATP levels increase. Thus the rate of citrate formation and hence fatty acid synthesis could be a function of the ATP levels in the cell. In this connection it is of interest to note that epinephrine increases the rate of fatty acid synthesis initiated by the addition of insulin and glucose to adipose tissue *in vitro.* One explanation that may be offered for this effect of epinephrine is that under these experimental conditions it sets in motion an ATP-consuming reaction involving the breakdown and resynthesis of triglycerides (see Chapter 8).

The other aspect of the citrate–malate cycle that is of interest is its generation of TPNH. In essence the cycle serves to convert extramitochondrial DPNH into TPNH. In so doing, it accomplishes two things. First, a supply of DPN^+ is maintained so that conversion of glucose to pyruvate can proceed rapidly. Second, it provides TPNH that is needed for the synthesis of fatty acids. It should be noted, however, that the amount of TPNH that can be generated by the citrate–malate cycle is only about half of the total amount needed for fatty acid synthesis. This is because only one DPNH is formed for each pyruvate molecule produced. Hence only one TPNH is formed by the citrate–malate cycle for each acetyl CoA unit that is generated. The formation of a fatty acid from acetyl CoA, however, requires slightly less than two TPNH molecules per acetyl CoA unit (for example, the conversion of eight acetyl CoA units into palmitic acid requires fourteen TPNH). This means that, for fatty acid synthesis to proceed, another source of TPNH must be furnished. This is achieved by the operation of a pathway in which glucose-6-phosphate is converted to a pentose phosphate. This pathway has been given numerous names – hexose monophosphate shunt, pentose cycle, or phosphogluconate oxidative pathway. We shall refer to it as the pentose cycle.

The first step in the pentose cycle is the oxidation of glucose-6-phosphate to 6-phosphogluconolactone with TPN^+ acting as an electron acceptor (cf. Fig. 7-2). The enzyme catalyzing this reaction is glucose-6-PO_4 dehydrogenase. The lactone produced is hydrolyzed by a lactonase to form 6-phosphogluconic acid, which serves to drive the reaction to the right and favors TPNH production. The second step in the cycle is the formation of another molecule of TPNH by the oxidation of the 6-phosphogluconic acid to the pentose, *D*-ribulose-5-PO_4. The 6-phosphogluconate dehydrogenase which catalyzes the reaction requires Mn^{++} as a cofactor. Decarboxylation occurs hand in hand with the dehydrogenation, and the reaction thus resembles that seen in the dehydrogenation of isocitrate in the citric acid cycle. It should be noted that the enzymes concerned in this oxidation of glucose-6-phosphate to the pentose are found in the cytosol. Hence here is the first case of CO_2 production outside the mitochondria by an oxidative process.

The D-ribulose-5-PO_4 produced is converted into two other pentoses, D-xylulose-5-PO_4 by the action of an epimerase and D-ribose-5-PO_4 by the action of an isomerase. The formation of these two pentoses leads to an interesting series of reactions which are depicted in Fig. 7-3. One molecule of xylulose-5-

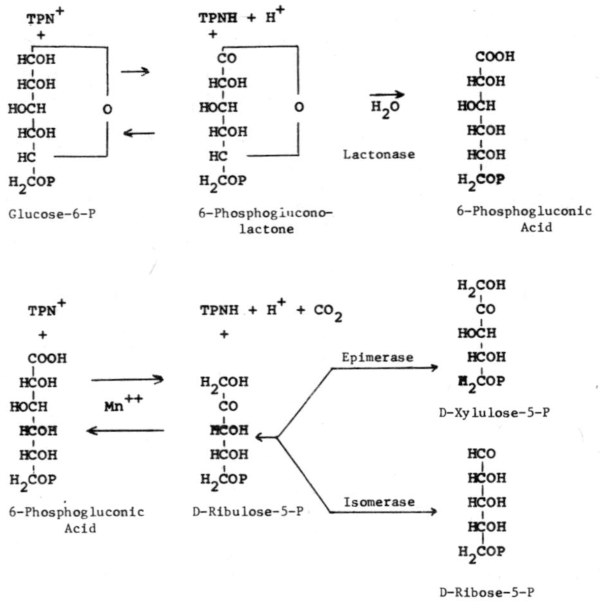

FIG. 7-2. The production of TPNH by the conversion of glucose-6-phosphate to D-ribose-5-phosphate in the pentose cycle.

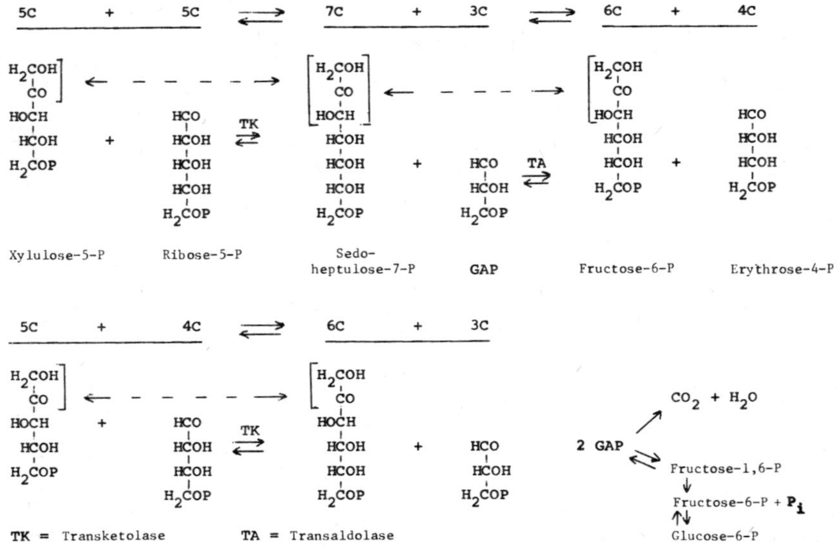

Sum:

$$6 \text{ Glucose} - 6\text{ P} + 12\ TPN^{+} \longrightarrow 5 \text{ Glucose} - 6\text{ P} + 12\ TPNH + 6\ CO_2 + 12\ H^{+}$$

FIG. 7-3. The conversion of pentose phosphates to hexose phosphates in the pentose cycle.

PO_4 and one of ribose-5-PO_4 react through the action of an enzyme named transketolase to form a seven-carbon sugar sedoheptulose-7-PO_4 and the three-carbon fragment, glyceraldehyde phosphate (GAP). These products in turn interact with a further exchange of carbons to form fructose-6-PO_4 and the four-carbon sugar, erythrose-4-PO_4. The enzyme catalyzing this reaction is named transaldolase. Now another molecule of the pentose, xylulose-5-PO_4 reacts with the erythrose-4-PO_4 formed to generate another molecule of fructose-6-PO_4 and glyceraldehyde phosphate, with tranketolase acting as the enzyme. The end result of this series of reactions is the converstion of three pentoses (fifteen carbons) into two molecules of fructose-6-PO_4 (twelve carbons) and one molecule of glyceraldehyde PO_4 (three carbons). Both end products will be recognized as intermediates in the main oxidative pathway of glucose discussed earlier. Thus the pentose PO_4 formed as a by-product of TPNH production is capable of being recycled through the pentose cycle or utilized for energy production. Reconversion of the pentose PO_4 to glucose-6-PO_4 can best be visualized by means of the following equations in which six molecules of glucose-6-PO_4 are employed.

$$6\ \text{Glucose-6-}PO_4 + 12\ TPN^+ \longrightarrow 6\ \text{Pentose-}PO_4 + 12\ TPNH^+$$

$$6\ \text{Pentose-}PO_4 \longrightarrow 4\ \text{Fructose-6-}PO_4 + 2\ \text{glyceraldehyde}\ PO_4$$

$$2\ \text{Glyceraldehyde}\ PO_4 \longrightarrow \text{Fructose-1,6-}PO_4$$

$$\text{Fructose-1,6-}PO_4 \longrightarrow \text{Fructose-6-}PO_4 + P_i$$

The fructose-6-PO_4 is readily converted to glucose-6-PO_4 by the enzyme phosphohexose isomerase (cf. Chapter 4). The overall reaction can thus be written as follows:

$$6\ \text{Glucose-6-}PO_4 + 12\ TPN^+ \longrightarrow 5\ \text{Glucose-6-}PO_4 + 12\ TPNH + 6\ CO_2 + 6\ H^+$$

Thus the complete oxidation of one molecule of glucose-6-PO_4 to CO_2 is accompanied by the production of 12 TPNH. If this TPNH was converted to DPNH by means of the following reaction catalyzed by the enzyme transhydrogenase:

$$TPNH + DPN^+ \longrightarrow DPNH + TPN^+$$

then twelve molecules of DPNH could be formed which when reacted with oxygen would yield 36 ∿P. It will be recalled that the oxidation of one molecule of glucose-6-PO_4 by the main oxidative pathway was calculated to yield 38 ∿P.

The reactions catalyzed by transketolase and transaldolase are reversible. This means that ribose-5-PO_4, a component of ribonucleic acid, can be synthesized from the glycolytic intermediates fructose-6-PO_4 and glyceraldehyde phosphate.

Transketolase requires Mg^{++} and thiamine pyrophosphate as a coenzyme to function. The thiamine pyrophosphate serves as a carrier of the two-carbon fragment that is transferred in the reaction. The intermediate formed may be formulated as follows:

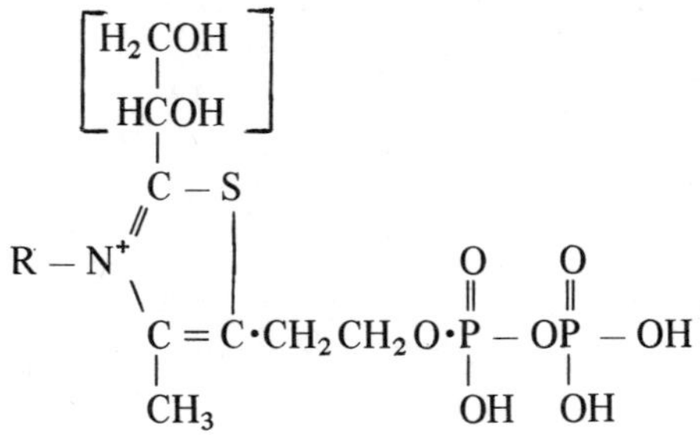

The action of thiamine pyrophosphate in this reaction is not unlike that encountered in the oxidation of pyruvate. Transaldolase functions by forming a Schiff base with the three-carbon fragment that is transferred. The NH_2 group of a lysyl residue in the enzyme is employed for this purpose, and the reaction can be represented as follows:

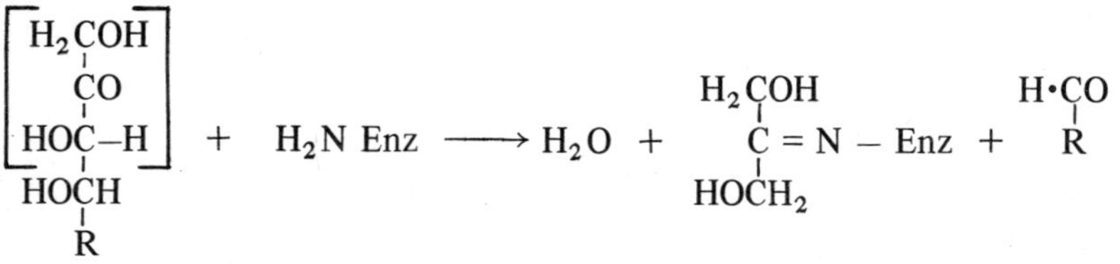

It should be noted that the action of transaldolase is similar to that of aldolase which cleaves fructose-1,6-diphosphate in the glycolytic cycle. The enzymes differ in that the Schiff base formed in the transaldolase reaction can only participate in an exchange reaction and is not cleaved by H_2O as in the case of the aldolase reaction.

The synthesis of the acyl carrier protein usually terminates when the carbon chain length reaches sixteen or eighteen. The acyl group is then apparently transferred from the carrier protein to the SH group of coenzyme A to form the corresponding acyl CoA. There then follows the formation of triglyceride. This occurs by the reaction of two acyl CoA units with glycerol phosphate to form a phosphatidic acid:

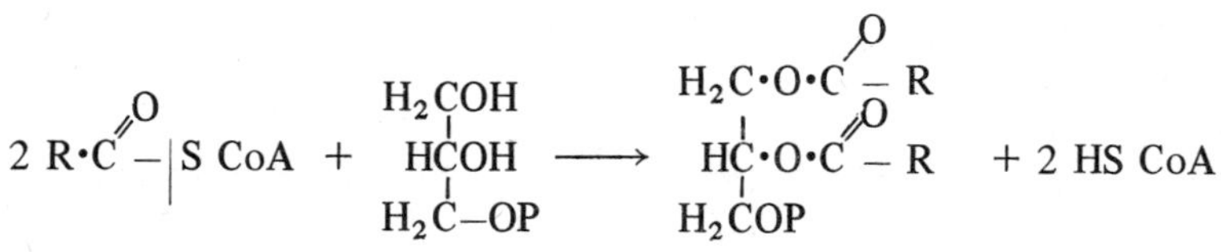

The phosphate group is then split off with water to form the diglyceride which can then react with a third acyl CoA unit to form the triglyceride:

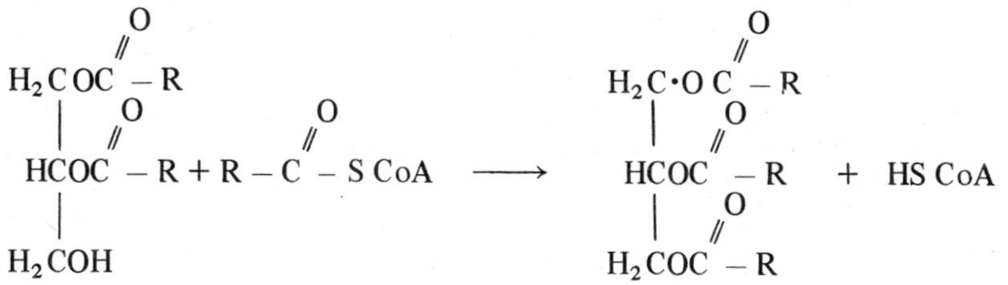

An alternative initial pathway for the formation of triglyceride exists in certain tissues. This involves the reaction of acyl CoA with dihydroxyacetone phosphate:

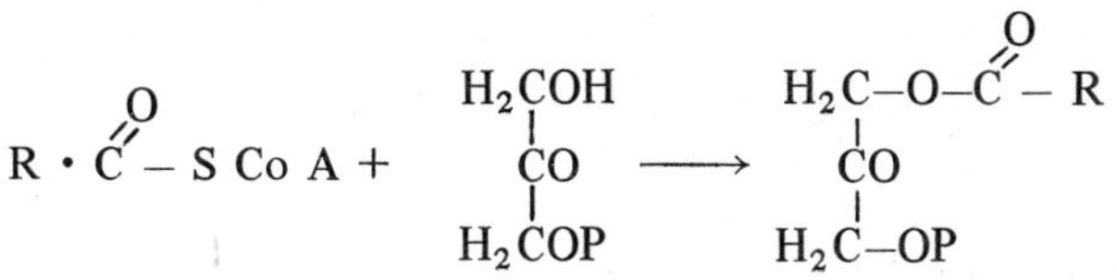

A reduction employing TPNH then occurs:

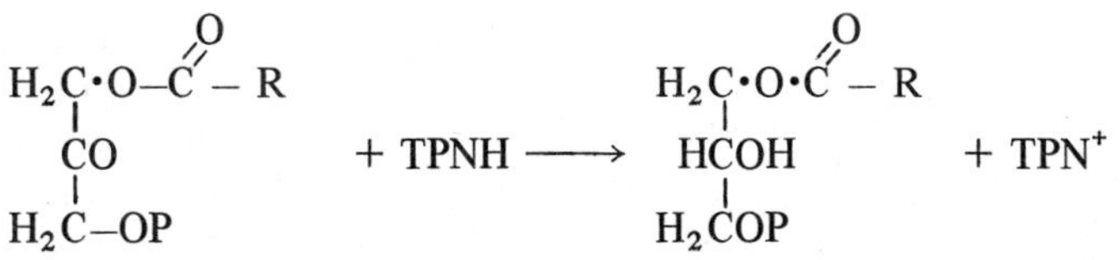

The compound formed can then go on to form phosphatidic acid.

The synthesis of fatty acid and the formation of triglyceride occur largely in the liver and adipose tissue. When the synthetic process occurs in adipose tissue, the triglyceride is deposited in the cell. The liver, however, does not normally retain the triglyceride that it synthesizes. The liver excretes its triglyceride into the bloodstream combined with a special protein to form a so-called lipoprotein. The triglyceride attached to this lipoprotein can be broken down in the capillaries of certain tissues by a lipoprotein lipase to yield glycerol and fatty acids. This lipase is activated by heparin. When cleavage of the triglyceride occurs in the adipose tissue capillaries, the fatty acids released may be picked up by the adipose tissue cells, reesterified to triglyceride, and deposited there.

CHAPTER 8

Energy Mobilization and Glucogenesis

The mobilization of the glycogen and fat stores of the animal for use as fuel in the production of energy presents one with a glimpse into a recent and interesting development in endocrinology. This is because the key agent in the activation of the enzymatic processes for the breakdown of both substances is cyclic AMP (cAMP). It is appropriate, therefore, that our discussion of energy mobilization begin with a consideration of the properties of this important compound.

Cyclic AMP has the following structure:

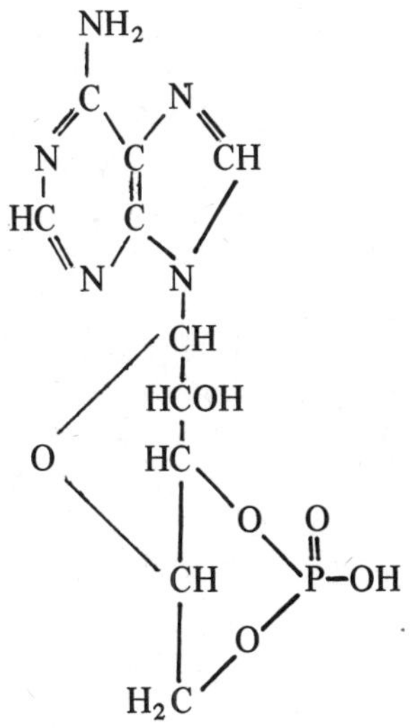

It is formed from ATP according to the following reaction:

$$\text{ATP} \xrightarrow{\text{Mg}^{++}} \text{cAMP} + \text{PP}_i$$

The enzyme catalyzing the reaction is called adenyl cyclase, and Mg^{++} is required as a cofactor. It has been reported that hydrolysis of cyclic AMP to AMP yields 14,000 calories per mole, an amount greater than that resulting from the cleavage of ATP to ADP and P_i. This should mean that little energy is dissipated in the formation of cyclic AMP and PP_i from ATP. Hydrolysis of the pyrophosphate to inorganic PO_4 by pyrophosphatase, which is present in the tissue, must then be essential if the equilibrium of the reaction is to be shifted in favor of cyclic AMP formation. The enzyme adenyl cyclase appears to be particulate in nature and to form part of the cell membrane.

Of prime importance is the fact that activation of adenyl cyclase and the production of cyclic AMP are brought about by hormones. A large variety of hormones have now been shown to act upon their specific target organs by eliciting the formation of cyclic AMP. The cyclic AMP so produced then, in turn, influences the activity of one or more enzyme reactions or cellular processes. Cyclic AMP has been termed a second messenger by Sutherland. It might also be called a primary or primitive regulator or hormone, since it is found in many lower forms and unicellular organisms. It is tempting to consider cyclic AMP as one of the primitive metabolic regulatory substances of the cell with its production in unicellular organisms subject to control by the availability of foodstuff in the environment as has been shown to be the case in *Escherichia coli*. One might then look upon the array of hormones found in man as an evolutionary development which has been, in large part, superimposed upon a basic cyclase system in order to permit its finer and specific regulation in individual tissues. The action of hormones on adenyl cyclase may be divided roughly into three categories. First, and most numerous, are those hormones that stimulate the production of cyclic AMP in their target tissues – for example, catecholamines, glucagon, ACTH, TSH, and vasopressin. A second type, of which insulin is an example, diminishes the concentration of cyclic AMP in the tissue. Some hormones such as prostaglandins or the catecholamines may, depending on the tissue involved, be placed in either the first or second category. Finally, there are those hormones that appear to have no direct and immediate action upon adenyl cyclase, but influence its sensitivity to those hormones in the first group. Thyroxine, growth hormone, and glucocorticoids may be cited as examples of hormones falling into this category. Other hormones could be cited as belonging to one or another of these groups, and the list is likely to grow in the future.

A common method by which cyclic AMP achieves its effect is by the activation of protein kinases which catalyze the following type of reaction:

$$\text{Protein} + \text{ATP} \longrightarrow \text{Phosphorylated protein} + \text{ADP}$$

Protein kinase

In a number of cases it has been shown that it is the hydroxy group of a seryl residue in the protein which becomes phosphorylated. Dephosphorylation of the protein is effected by a phosphatase, and evidence is accumulating that this enzyme may also be under hormonal control. Proteins such as casein, protamine, and histone have been shown to serve as substrates for cyclic AMP-stimulated protein kinases and have been utilized to measure cyclic AMP production and the activation process. It has also been suggested that the induction of enzyme synthesis in certain bacteria and in animal cells by cyclic AMP may be brought about by the ability of cyclic AMP to activate the phosphorylation of proteins. Envisaged in such an action is a change in DNA–protein interactions as the protein becomes phosphorylated, which leads to an induction of RNA and protein synthesis. When the protein undergoing phosphorylation is an enzyme, then the phosphorylated product may be more or less active than its precursor. In such cases the end result of cyclic AMP action is either to slow down or to accelerate certain metabolic processes and usually in such a way that catabolic reactions are favored. Examples of this will be seen later in this chapter in the presentation of glycogen and triglyceride breakdown.

Some insight into the manner in which cyclic AMP activates a protein kinase is beginning to emerge. It appears that a protein kinase exists in the tissue combined with a smaller protein fragment which renders the kinase inactive. Cyclic AMP apparently is able to cause this inhibitor to dissociate from the kinase and thus permit it to function. It has also been suggested that cyclic AMP may influence a kinase by an allosteric effect – for example, phosphofructose kinase.

As with all hormones, a mechanism for terminating the action of cyclic AMP exists. This is achieved by the action of a specific phosphodiesterase which brings about the hydrolysis of cyclic AMP to AMP. This phosphodiesterase is inhibited by the methyl xanthines – for example, caffein and theophylline. Hence these substances permit cyclic AMP to accumulate and enhance or mimic the action of hormones which stimulate cyclic AMP production. The methyl xanthines have therefore proved to be useful compounds in elucidating the role of cyclic AMP in the control of various cellular metabolic processes.

With this general introduction to the role of cyclic AMP, we may now turn to a specific instance of its action in the control of glycogen metabolism. Here the details of the mechanisms involved are well documented and serve admirably to portray the ingenious control of metabolic processes that is afforded by cyclic AMP production.

The basic reaction of glycogen breakdown may be expressed as follows:

$$\text{Glycogen } (n \text{ glucose residues}) + P_i \longrightarrow \text{Glycogen } (n-1 \text{ glucose residues}) + \text{glucose-1-}PO_4$$

The enzyme catalyzing this reaction is called glycogen phosphorylase and acts upon the α-glucosidic-1,4 linkages of glycogen. The phosphorylase exists in two forms, one of which is phosphorylated. This is referred to as phosphorylase a and can be considered as the active form of the enzyme in the cell. The unphosphorylated enzyme is called phosphorylase b. The phosphorylation reaction as carried out by phosphorylase kinase in muscle may be represented as follows:

$$\text{Phosphorylase b} + \text{ATP} \longrightarrow \text{Phosphorylase a} + \text{ADP}$$

Both the b and a forms exist as dimers, and the a form has been reported to also exist as a tetramer. The molecular weight of the monomeric form is approximately 90,000. Phosphorylation has been shown to take place on the hydroxy group of a seryl residue in the enzyme. Each monomer contains one molecule of pyridoxal phosphate, and its removal from the enzyme results in inactivation. Activity is restored by the addition of this coenzyme. The mode of action of pyridoxal phosphate in the enzymatic process remains to be elucidated. Conversion and inactivation of the phosphorylated enzyme, form a, to the unphosphorylated state, form b, is brought about by a phosphatase.

Phosphorylase b kinase, the enzyme that converts form b to a, also exists in two states, an active phosphorylated species and an inactive unphosphorylated species. The conversion of one form to the other:

$$\text{Phosphorylase b kinase} + \text{ATP} \longrightarrow \text{Phosphorylase b kinase } PO_4 + \text{ADP}$$

is brought about by a protein kinase which is activated by cyclic AMP. Inactivation of phosphorylase b kinase is achieved by the action of a phosphatase. The reason for this involved control mechanism of phosphorylase is not clear, but it should be noted that there is good evidence to indicate that the same protein kinase that acts upon phosphorylase b also is able to carry out the phosphorylation of glycogen synthetase.

The formation of glycogen and its breakdown may thus be represented in outline form as follows, where the enzymes involved are represented by the letter E:

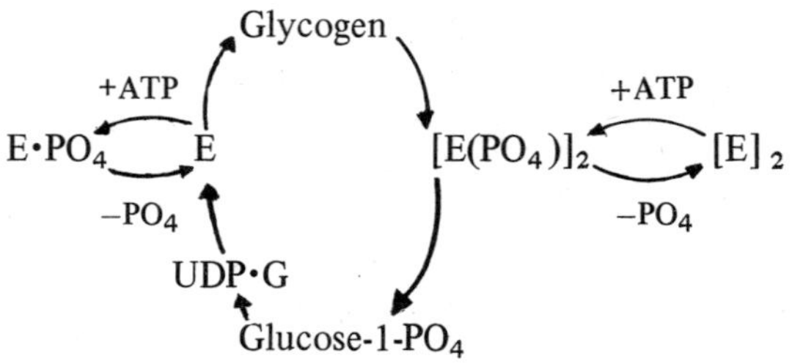

The production of cyclic AMP acting through the same protein kinase thus serves to activate the phosphorylation of the enzymes involved in both the synthesis and breakdown of glycogen. Thus the synthetic pathway is shut down and the mobilization of glycogen stores is turned on. There is thereby achieved the prevention of an energy-wasteful process in which glucose-1-PO_4 could be continuously recycled through glycogen. A useful mnemonic that may be helpful in memorizing this chain of events is POGO: Phosphate On (the enzyme), Glucose Off (glycogen); Phosphate Off (the enzyme), Glucose On (glycogen).

It will be recalled that liver and muscle are the two tissues in which appreciable amounts of glycogen are stored. Although the reactions for mobilization of glycogen in these two tissues are basically similar, there are important differences that need emphasis. First is the hormonal control that activates the production of cyclic AMP. In muscle the adenyl cyclase is activated only by epinephrine, while in liver both epinephrine and glucagon perform this function, though glucagon seems to be the more potent activator in this tissue. Second, the breakdown of glycogen in the two tissues yields different end products in so far as energy utilization of glycogen stores by the whole organism is concerned. In both tissues the glucose-1-PO_4 formed can be converted to glucose-6-PO_4. In liver, but not in muscle, there exists a phosphatase that converts glucose-6-phosphate to glucose. Thus in liver a release into the bloodstream of glucose formed from glycogen is possible. In muscle the glucose-6-phosphate formed can contribute to a supply of energy for other tissues only if it has been converted to lactate. The lactate released from muscle into the bloodstream can be utilized as such by other tissues, or it may be converted to glucose in the liver by the process of glucogenesis which will be discussed shortly.

The mobilization of the fat stores of the body involves a process not unlike that for glycogen. However, our knowledge of the enzymes which carry out the conversion of triglyceride to fatty acids and glycerol is less complete than for those involved in glycogen breakdown. This is partly because investigations in this area are of a more recent origin. Most of our information about the mobilization of the triglyceride stores of adipose tissue is the result of studies on the rat, and these studies will form the main basis of our presentation. Evidence is available, however, to indicate that considerable differences may be encountered among different species, especially as regards the details of hormonal control.

Adipose tissue contains a lipase which can be converted from an inactive form to an active one by hormonal action. There is evidence to indicate that the active form of this lipase brings about the hydrolysis of triglyceride to diglyceride and fatty acid. The further breakdown of the diglyceride to fatty acids and glycerol then appears to occur by other lipases present in the tissue which require no activation. Activation of the lipase specific for triglyceride in rat adipose tissue is brought about by a large number of hormones which include glucagon, TSH, ACTH, and the catecholamines. All these hormones induce the formation of cyclic AMP in adipose tissue. The specificity of cyclic AMP production by a single hormone that is seen in other tissues is thus lacking at least in rat adipose tissue. The reason for this is not clear, though one could speculate that mobilization of energy stores is a desirable accompaniment of the action of each of these hormones on its specific target tissue. There is evidence to indicate that the adenyl cyclase of adipose tissue has a separate receptor site for each of the hormones that activates it. There seems little doubt that the production of cyclic AMP is the key to hormonal activation of the triglyceride lipase. For example, strong activation of this lipase can be achieved by the addition to adipose tissue of theophylline or caffein, inhibitors of the phosphodiesterase which converts cyclic AMP into AMP. The dibutyryl derivative of cyclic AMP which penetrates cell membranes more readily than the parent compound also evokes lipase activation. The mode of action of cyclic AMP appears to conform to the basic pattern outlined earlier in which it activates a protein kinase which then catalyzes the following reaction:

$$\text{Lipase} + \text{ATP} \longrightarrow \text{Phosphorylated lipase} + \text{ADP}$$

the phosphorylated form of the enzyme being active. Conversion of the active form of the lipase to the inactive form is then presumably achieved by a phosphatase, though the evidence for this requires further documentation.

The end result of the action of hormones upon adipose tissue is the production of fatty acids, referred to in the literature as free fatty acids (FFA) or nonesterfied fatty acids (NEFA), and glycerol. The production of glycerol serves as the best index of the lipolysis rate, since this compound, unlike the fatty acids, is not readily utilized by adipose tissue, owing to its negligible content of glycerol kinase. *In vivo,* most of the glycerol is released into the bloodstream to be picked up by tissues such as the liver, which can readily convert it to glycerol phosphate. The FFA formed may also be released to the bloodstream where it is transported to other tissues in combination with the serum albumin. It must be noted, however, that the FFA produced may undergo oxidation to supply the basic energy needs of adipose tissue. More important, however, is the fact that adipose tissue is capable of carrying out the reesterification of FFA to re-form triglyceride. It can perform this operation, however, only if glucose is available to it for the formation of glycerophosphate, and in this connection it should be noted that the entrance of glucose into adipose tissue is markedly dependent upon the presence of insulin. *In vitro* it has been shown that the addition of epinephrine and insulin to adipose tissue in the presenee of glucose results in a three- to fourfold increase in oxygen consumption with the release of glycerol, but no production of FFA. These results have been interpreted to indicate that lipolysis is stimulated with the reesterification of the fatty acid back to triglyceride. The increase in oxygen consumption is the result of the need for high-energy phosphate for the reesterification process (see Chapter 7). Since the overall process is in essence wasteful of energy and must result in the generation of heat, it has been proposed that such a process might be employed by an animal to maintain body temperature when subjected to adverse environmental surroundings.

The importance of serum albumin for the release of FFA from adipose tissue is demonstrated by *in vitro* studies. If epinephrine is added to adipose tissue incubated *in vitro* in an isotonic buffered medium, FFA accumulates in the tissue and none appears in the medium. Tissue respiration gradually fails, and lipolysis stops. If serum albumin is present in the medium, then FFA are released from the tissue into the medium. Respiratory and lipolysis rates then proceed normally until the capacity of the serum albumin to bind fatty acid is exceeded. Each serum albumin has a number of binding sites for fatty acid. Two of these binding sites have a high affinity for fatty acids with association constants of the order 10^7 to 10^8. The constants for the remaining binding sites are several magnitudes lower. At the normal concentration of albumin in the serum it can be calculated that the two high-affinity sites become saturated when the serum level of FFA approaches a value of approximately 1000 μeq/liter. Normal values in humans for serum FFA are in the range of 400 to 800 μeg/liter. Values that exceed 1000 μeq are seen, for example, in diabetes, hyperthyroidism, myocardial infarction, and starvation. The concentration of unbound or more loosely bound serum FFA in such conditions must rise. Since fatty acids are known to be potent uncouplers of oxidative phosphorylation and inhibitors of numerous enzymatic processes, the question of the toxic effects of fatty acids in such disorders has been raised.

The increase in serum FFA in diabetes is worthy of brief consideration, since it serves to illustrate the main aspects of the control mechanism involved in FFA mobilization and metabolism. It has long been known that blood glucose rises in the diabetic state, but the concomitant increase in serum FFA that occurs has only been recognized more recently. Figure 8-1 serves to illustrate the concomitant changes in serum glucose and FFA following the injection of insulin into a normal human. The fall in FFA values slightly precedes that of glucose. Similar changes after insulin injection are seen in the diabetic where both serum glucose and FFA may be two or three times their normal value. Two reasons can be evoked for the effect of insulin on serum FFA concentration. First, insulin exerts an antilipolytic effect on adipose tissue. This may be seen very clearly in studies on adipose tissue when it is incubated *in vitro* in the absence of glucose. The addition of epinephrine to such tissue induces a marked increase in the production of fatty acids and glycerol. If insulin is also present, the lipolytic response to epinephrine will be markedly diminished. This effect of insulin is best seen when the concentration of epinephrine does not exceed 0.1 μg/ml or is near the physiological level. Under such conditions the antilipolytic action of insulin can be detected at concentrations (1 to 10 μ units/ml) which are equal to or even slightly less than those necessary for the demonstration of an effect on glucose uptake by adipose tissue. The antilipolytic action of insulin increases as its concentration is increased, and a 90% inhibition of lipolysis may be achieved at insulin levels of 100 to 1000 μ units/ml. If the epinephrine concentration is raised to values in the range of 5 to 10 μg/ml, then little antilipolytic action of insulin is seen, no matter how high the insulin concentration is made. Adipose tissue taken from a diabetic rat displays a spontaneous high lipolysis rate, and this can be markedly reduced by the addition of insulin to the incubation medium. Thus this antilipolytic action of insulin may be one reason for the fall seen in serum FFA after insulin injection. Whether it can account for the fall in FFA values prior to those seen in glucose concentration, as illustrated in Fig. 8-1, remains unproved. The antilipolytic action of insulin is accompanied by a fall in tissue cyclic AMP levels. The evidence available suggests that insulin in some unknown way either diminishes the activity of adenyl cyclase, or increases the rate of destruction of cyclic AMP by activating phosphodiesterase.

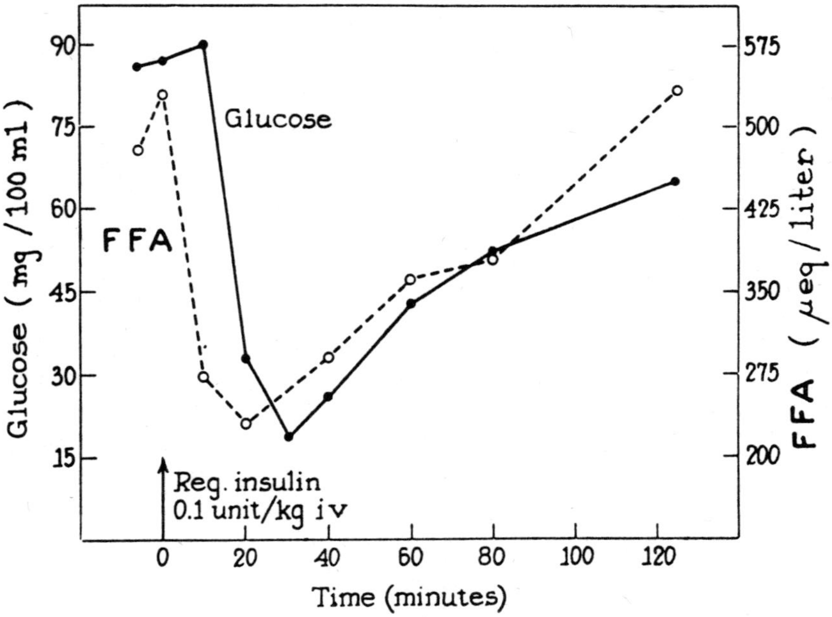

FIG. 8-1. The effect of insulin injection on the plasma glucose and free fatty acids (FFA). At the indicated time, insulin (regular) was injected intravenously into a normal human subject. One unit of insulin equals 0.04 mg. (Reproduced from V.D. Dole, *in* I.H. Page (ed.), *Chemistry of Lipids as Related to Atherosclerosis,* Charles C Thomas, Springfield, Illinois, 1958).

The second reason that can be offered for the effect of insulin on serum FFA levels is related to the ability of insulin to facilitate the passage of glucose into adipose tissue. As was mentioned earlier in this chapter, adipose tissue has a high capacity to carry out the esterification of fatty acids to triglyceride. However, it can do this only if glucose is available to furnish the glycerophosphate needed for triglyceride formation (see Chapter 7). Thus the release of FFA by adipose tissue may be curtailed by insulin *in vivo* owing to a substantial portion of the FFA formed by lipolysis being converted back to triglyceride. In addition, when the serum FFA levels are high, adipose tissue, when it is supplied with glucose, may absorb FFA from the bloodstream and convert it to triglyceride.

Some of these aspects of fatty acid mobilization and metabolism are depicted in Fig. 8-2. The rise in serum FFA that occurs after a starvation period of 2 to 3 days or longer would appear to be triggered by a different mechanism than that

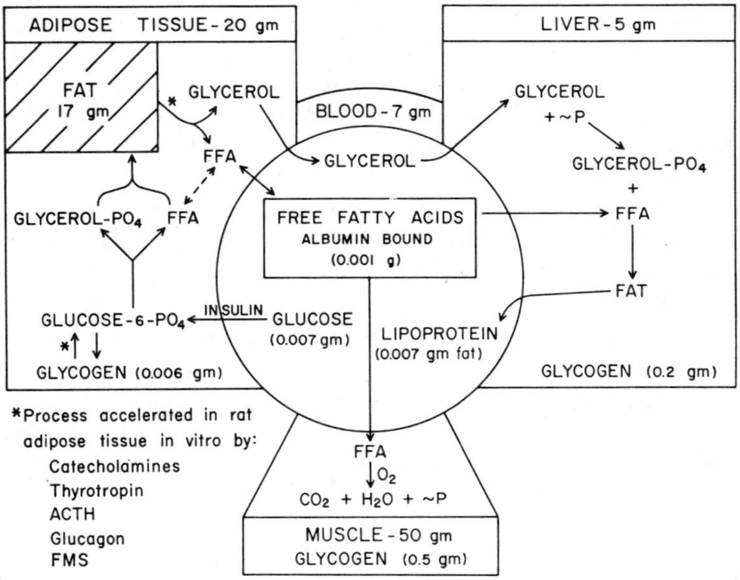

FIG. 8-2. Interrelationships in the metabolism of free fatty acids (FFA) by adipose tissue, liver, muscle, and blood. The weight given for each tissue represents the average amount of tissue found in 100 g of a well-fed animal. The amounts of glucose, glycogen, FFA, and fat are average values which serve as an index to the relative importance of the various energy sources of the animal. For example, if in the rat the average O_2 consumption is taken as 2 cc per 100 g of animal per minute, then the time required for the different tissue constituents to be consumed may be calculated. These are as follows: blood–FFA, 1 minute; glucose, 2.4 minutes; triglceride in lipoprotein, 6 minutes; liver glycogen, 1.2 hours; muscle glycogen 3.1 hours; adipose tissue triglyceride, 254 hours. In man, where the rate of oxygen consumption is approximately one-fifth that of the rat, these amounts of tissue constituents should last five times as long. [Reproduced from E.G. Ball and R.L. Jungas, *Recent Prog. Hormone Res.* **20,** 183–214 (1964.]

seen during more normal mobilization of fat stores or in diabetes. For example, insulin has no antilipolytic action on the increased lipolysis rate observed *in vitro* on adipose tissue taken from starved rats. The appearance in the urine of starved animals of a substance called fat-mobilizing substances (FMS) has been reported by numerous workers. The adipose tissue of rats fasted for 3 days and then refed a high carbohydrate diet presents an interesting metabolic pattern when studied *in vitro*. The high lipolysis rate due to starvation persists as shown by profuse glycerol production, but FFA production is near zero. This type of tissue possesses very high stores of glycogen which can be shown to break down on incubation *in vitro* to furnish glycerophosphate for reesterification of the released FFA. If insulin is now added to this tissue, it exerts a marked antilipolytic action, in contrast to its failure to act on the lipolytic process in starved tissue. The antilipolytic action of insulin displayed on tissue from fasted and refed rats is manifested by a marked diminution of glycerol production and a fall in the rate of oxygen consumption. Concurrently an interesting and unexplained increase occurs in the *de novo* synthesis of fatty acids from the glycogen stores of this tissue. It is obvious that some interesting facets of the process of fat mobilization and insulin action await elucidation.

The final process under the heading of energy mobilization to be considered is the formation of glucose from substances other than glycogen. This process is termed glucogenesis, or, when it entails the conversion of pyruvate or oxalacetate derived from amino acids to glucose, the term gluconeogenesis is often employed. Lactate derived from muscle glycogen or glycerol derived from triglyceride can be important precursors of glucose. Gluconeogenesis or glucogenesis, though not ranking quantitatively with triglyceride and glycogen as a means for supplying energy demands, can be important as a source of glucose when glycogen stores are depleted. A key reaction in the production of glucose from metabolites which can be converted to oxalacetate is the formation of phosphoenol pyruvate (PEP). This occurs by the following reaction catalyzed by PEP carboxykinase:

$$\begin{array}{lcl} \mathrm{COOH} & & \\ | & & \\ \mathrm{CH_2} & + \ \mathrm{GTP} \longrightarrow & \mathrm{CH_2} \\ | & & \| \\ \mathrm{CO} & & \mathrm{CO{\sim}PO_3H_2} + \mathrm{CO_2} + \mathrm{GDP} \\ | & & | \\ \mathrm{COOH} & & \mathrm{COOH} \end{array}$$

This is one of the few reactions in which ATP does not furnish directly the high-energy phosphate that is required. Guanosine triphosphate (GTP) may, of course, be formed from GDP by reaction with ATP. If pyruvate derived from amino acids or lactate is to enter the glucogenic pathway, it must first be converted to oxalacetate. This can be accomplished by the pyruvate carboxylase reaction:

$$\text{Pyruvate} + CO_2 + \text{ATP} \longrightarrow \text{Oxalacetate} + \text{ADP} + P_i$$

which was discussed in connection with fatty acid synthesis (Chapter 7).

Once phosphoenol pyruvate is formed, it may be converted to fructose diphosphate by a reversal of the steps used in glycolysis (Chapter 3):

Phosphoenol pyruvate
↓
2-Phosphoglycerate
↓
3-Phosphoglycerate
↓ + ATP
1,3-Diphosphoglycerate
↓ + DPNH
Triose phosphate
↓
Fructose-1,6-diphosphate

Tissues that are capable of carrying out glucogenesis, such as liver and kidney, contain two phosphatases. One, acts on fructose-1,6-diphosphate to form fructose-6-phosphate. This product is converted by phosphohexoisomerase to glucose-6-phosphate, which in turn can be acted upon by glucose-6-PO_4 phosphatase to yield glucose. The overall process in which pyruvate is converted to glucose can be written as follows:

$$2\ \text{Pyruvate} + 6{\sim}P + 2\ \text{DPNH} \longrightarrow \text{Glucose}$$

Since it will be recalled that the reverse process is

$$\text{Glucose} \longrightarrow 2\ \text{Pyruvate} + 2{\sim}P + 2\ \text{DPNH}$$

gluconeogenesis is an energy-consuming process. This is important to emphasize, since the glucogenic pathway is most active when the body energy stores, especially glycogen, have been subjected to some drain, such as starvation. Hence the need for glucose for special purposes such as brain metabolism has been suggested as the determining factor for the operation of this pathway.

The operation of the glucogenic pathway requires the passage of metabolites in a reverse direction along part of the main highway of glucose metabolism. There thus arises the question of traffic control along this highway so that the desired direction of flow is achieved. To be sure, as in the case of starvation when the glucogenic pathway is operative, there is little traffic in the direction of glucose to pyruvate, so that a reverse flow is favored. Though this is helpful, it alone is not enough. Control mechanisms are also needed to prevent what may be termed diversionary or futile reactions. The limited presentation that follows will serve to indicate the chief points in this area that need consideration.

If we start with pyruvate as a potential source of glucose, the first point that requires consideration is the obvious need to prevent conversion of pyruvate to acetyl CoA, so that the limited amount of pyruvate available will be channeled into the formation of oxalacetate and PEP. Recent work indicates that the pyruvate dehydrogenase complex can exist in a phosphorylated and an unphosphorylated form. Conversion to the inactive or phosphorylated form is inhibited by insulin and stimulated by epinephrine. Thus one may draw the conclusion that in starvation, when blood levels of glucose and consequently insulin are low, pyruvate oxidation would be impaired. Conversion of pyruvate to oxalacetate by the action of pyruvate carboxylase should then be favored, especially since this enzyme is activated by acetyl CoA, a metabolite that should be plentiful during starvation, when metabolism of fatty acids is predominant.

If we accept the premise that any cell containing PEP carboxykinase and given a supply of oxalacetate will produce PEP, then the next major problem that arises is the control of the pathway followed by PEP. It will be remembered that, in the glycolytic cycle, pyruvate kinase catalyzes the transfer of high-energy phosphate from PEP to ADP with the formation of ATP and pyruvate. Obviously this reaction must be prevented from occurring; otherwise destruction of PEP results, energy is wasted, and a futile cycle – from pyruvate to pyruvate – is produced. Some manner of control of the activity of pyruvate kinase would seem to be required for glucogenesis to proceed. It would appear that the pyruvate kinase of a glucogenic tissue such as liver is different from that found, for example, in muscle. The activity of the enzyme in liver, but not in muscle, is reported to be lowered in starvation and in diabetes. Administration of insulin to diabetic animals restores activity. It has also been reported that ATP and alanine inhibit the pyruvate kinase of rat liver.

The conversion of PEP to fructose diphosphate in the glucogenic cycle presents no special problems of the type discussed. However, the portion of the pathway that lies between fructose diphosphate and glucose presents control problems. First, the conversion of fructose diphosphate to fructose-6-phosphate requires that fructose diphosphatase be active and that phosphofructokinase be inactive, if a futile cycle is to be prevented. The reverse situation must ensue when glucose oxidation is to proceed. It has been proposed that the AMP level in the cell may be an important factor in such a control.

This nucleotide has been reported to stimulate the activity of phosphofructokinase and to inhibit fructose diphosphatase. A similar situation exists in the consideration of the enzymes, hexokinase and glucose-6-phosphatase, concerned in the reactions whereby the interconversion of glucose and glucose-6-phosphate occurs. An increase in the activity of liver glucose-6-phosphatase has been reported to occur during starvation and diabetes, or after the administration of steroid hormones. A fuller discussion of the problems encountered in the regulation of glycolysis and glucogenesis may be found in the review by Scrutton and Utter.

All the enzymes needed to convert PEP to glucose are found in the cytoplasm. The two enzymes needed to convert pyruvate to PEP – pyruvate carboxylase and PEP carboxykinase – have been reported to be present in both the mitochondria and the cytosol. A marked species difference occurs in the location of PEP carboxykinase; in rat liver it is found predominantly in the cytosol; in guinea pig liver it is found in both the cytosol and mitochondria; and in pigeon liver it is predominantly a mitochondrial enzyme. In those species where pyruvate carboxylase is mainly a mitochondrial enzyme, but PEP carboxykinase is in the cytosol, consideration needs to be given to the mode of transport of oxalacetate across the mitochondrial membrane. It has been proposed that transport of oxalacetate is achieved by its conversion in the mitochondria to aspartate or malate, substances to which the mitochondrial membrane is apparently more permeable. The use of malate for this purpose has the added advantage that, when it is reconverted to oxalacetate by the action of malate dehydrogenase, generation of extramitochondrial DPNH also occurs. This transfer of reducing equivalent from the mitochondria could thus furnish the DPNH needed for the reduction of 1,3-diphosphoglycerate. Presumably when lactate is the predominant substrate for glucogenesis, such a transfer of reducing equivalents would be unnecessary.

GLOSSARY

ACP	Acyl carrier protein. A protein which binds acyl intermediates during the synthesis of fatty acids.
ACTH	Adrenocorticotropic hormone. It is secreted by the hypophysis and stimulates the synthesis and secretion of adrenal cortical steroids.
ADP	Adenosine diphosphate.
AMP	Adenosine monophosphate, adenylic acid.
Anaerobic	Occurring in the absence of oxygen.
Antibiotic	A compound usually produced by one living organism which kills or inhibits the growth of other organisms.
ATP	Adenosine triphosphate.
cAMP	Cyclic AMP, 3′,5′-cyclic adenylic acid.
Catecholamines	Epinephrine and norepinephrine.
Chelate	A coordination complex usually formed between a metallic ion and one or more groups of an organic compound.
Chemical energy	The potential energy of a chemical compound that may be converted into other forms of energy – for example, free energy, heat.
CoA, CoASH	Coenzyme A.
Coenzyme	A comparatively low-molecular-weight organic substance which can attach itself to specific proteins to form an active enzyme system.
Combustion	The chemical reaction of oxygen with a fuel and accompanied by the liberation of useful energy and heat.
Cristae	The series of infoldings formed by the inner membrane of mitochondria, which project to various depths into the cavity of the mitochondria.

Cytoplasm	Usually defined as the material of the cell exclusive of the nucleus, which is surrounded by the cell membrane.
Decarboxylation	The removal of CO_2 from the carboxyl group, COOH, of a substrate. A biological process catalyzed by an enzyme called a decarboxylase.
Dehydration	The removal of H_2O from a substrate.
Dehydrogenase	An enzyme that catalyzes the process of dehydrogenation.
Dehydrogenation	An oxidation process in which hydrogen ($H^+ + e$) is removed from a substrate.
DPN^+	Diphosphopyridine nucleotide.
DPNH	The reduced form of DPN^+
e	Electron, the most elementary charge of negative electricity.
E.T.S.	Electron transmitter system. A series of oxidation–reduction systems which carry electrons from a primary electron acceptor to oxygen.
Electrophoresis	The migration of a molecule with a net positive or negative charge in an electric field.
Energy	The capacity for doing work. That property of a system which diminishes when the system does work on any other system.
Exergonic process	A process in which a decline in free energy occurs.
FAD	Flavin adenine dinucleotide.
FFA	Free fatty acid, a nonesterified fatty acid (NEFA).
FP	Flavoprotein.
Free energy	The maximum work obtainable at constant temperature in a reaction.
Free radical	An atom or group of atoms such that there is at least one unpaired electron.
Glucogenesis	The formation of glucose from pyruvate or oxalacetate and substances (amino acids) which give rise to these compounds.

Glutathione	The tripeptide, γ-glutamylcysteinylglycine.
Glycogen	A branch-chained polysaccharide of glucose, a counterpart of starch and the storage form of glucose in animals.
Glycolysis	The conversion of glucose to lactic acid, a process which can proceed under anaerobic conditions.
GTP	Guanosine triphosphate.
High-energy bond	A term loosely employed to describe a linkage in a chemical compound which will yield a relatively large amount of free energy upon hydrolysis.
High energy phosphate	A high-energy bond in which the linkage involves a derivative of phosphoric acid.
Hormone	A substance that regulates the rate of a cellular process. Hormones are commonly formed in one tissue and transported by the blood to their site of action.
Isoelectric point	The pH at which the net positive and negative charges of a molecule are equal and no migration occurs in an electric field (electrophoresis).
Kcal	Kilocalorie (1000 calories). One calorie is the amount of heat required to raise the temperature of 1 g of water from 15°C to 16°C.
Kinase	An enzyme that catalyzes the transfer of certain groups from one molecule to another. Most usually ATP is one of the molecules involved.
K_m	The Michaelis or Michaelis–Menten constant. A measure of the relative affinity of a substrate for an enzyme.
Make-ready reaction	A term applied to a reaction in which a substrate undergoes an alteration in preparation for a dehydrogenation.
Matrix	The material lying within the confines of the inner membrane of a mitochondria.
Metabolism	The chemical processes in a living cell involved in the transformation of one substance into another and usually accompanied by a change in free energy.

Mitochondria	An organelle found in the cytoplasm of the cell which is surrounded by a double membrane, the innermost forming cristae. It carries out the major reactions of biological oxidations.
Nucleotide	A compound containing a heterocylic base (purine, pyrimidine, etc.), a pentose, and phosphate.
Oxidation	A process in which electrons are removed from a compound.
Oxidation–reduction potential	The voltage difference at a fixed temperature and pH between a solution containing equal amounts of an oxidant and reductant of an oxidation–reduction system and some agreed-upon standard oxidation–reduction system.
Oxidation–reduction system	A mixture of the reduced and oxidized forms of a substance. The interconversion of one form to the other involves the addition or removal of an electron or electrons and is readily reversible.
Oxidative phosphorylation	The formation of high-energy phosphate (ATP) which accompanies the passage of electrons to oxygen by way of the electron transmitter system.
$\sim$P	High-energy phosphate.
PEP	Phosphoenol pyruvate.
Phospholipid	A group of lipid compounds that yield on hydrolysis phosphoric acid, an alcohol, fatty acid, and a nitrogenous base – for example, lecithin.
P_i	Inorganic phosphate.
PP_i	Pyrophosphate.
Primary electron acceptor	The acceptor of electrons in the dehydrogenation of a metabolite.
Receptor site	The point of attachment of a hormone on or at the cellular component that it affects.
Schiff base	A compound formed by the reaction of an aldehyde or ketone with primary amines.
Seryl residue	The grouping $-NH \cdot CH_2OH \cdot CO-$ formed when serine is incorporated into a protein.

Substrate	A compound that is subject to alteration by the action of a specific enzyme.
TPN^+	Triphosphopyridine nucleotide.
TPNH	The reduced form of TPN^+.
TSH	Thyroid-stimulating hormone, or thyrotropin. The hormone released by the hypophysis which stimulates the production of thyroxine by the thyroid gland.
UTP	Uridine triphosphate.
Vitamin	An organic substance essential for the normal functioning of an organism, which must be supplied in the diet.

REFERENCES

GENERAL – Textbooks of Biochemistry

LEHNINGER, A. L., *Biochemistry,* Worth, New York, 1970.

McGILVERY, R. W., *Biochemistry, A Functional Approach,* W. B. Saunders Co., Philadelphia, 1970.

WHITE, A., HANDLER, P., and SMITH, E. L., *Principles of Biochemistry,* 4th ed., McGraw-Hill Book Company, New York, 1968.

CHAPTER 1

CLARK, W. M.. *Oxidation Reduction Potentials of Organic Systems,* Williams & Wilkins Co., Baltimore, 1960.

CHAPTER 2

GREVILLE, G. D., and TUBBS, P. K., The Catabolism of Long Chain Fatty Acids in Mammalian Tissues *in* P. N. Campbell and G. D. Greville (eds.), *Essays in Biochemistry* Vol. 4, Academic Press, New York, 1968.

LOWENSTEIN, J. M., The Pyruvate Dehydrogenase Complex and the Citric Acid Cycle, *in* M. Florkin and E. H. Stotz (eds.), *Comprehensive Biochemistry,* Vol. 18 S, pp. 1–55, Elsevier Publishing Co., New York, 1971.

REED, L. J., Chemistry and Function of Lipoic Acid, *in* M. Florkin and E. H. Stotz (eds.), *Comprehensive Biochemistry,* Vol. 14, pp. 99–126, Elsevier Publishing Co., New York, 1967.

WAKIL, S. J., and BARNES, E. M., Jr., Fatty Acid Metabolism, *in* M. Florkin and E. H. Stotz (eds.), *Comprehensive Biochemistry,* Vol. 18 S, pp. 56–104, Elsevier Publishing Co., New York, 1971.

CHAPTER 3

AXELROD, B., Glycolysis, *in* D. M. Greenberg (ed.), *Metabolic Pathways,* 3rd ed., Vol. 1, pp. 112–145, Academic Press, New York, 1967.

DICKENS, F., RANDLE, P. J., and WHELAN, W. J. (eds.), *Carbohydrate Metabolism and Its Disorders,* Vols. I and II, Academic Press, New York, 1968.

REED, L. J., Pyruvate Dehydrogenase Complex, *in* B. L. Horecker and E. R. Stadtman (eds.), *Current Topics in Cellular Regulation,* Vol. 1, Academic Press, New York 1969.

CHAPTER 4

CHANCE, B., and WILLIAMS, G. R., The Respiratory Chain and Oxidative Phosphorylation, *Advan. Enzymol.* 17, 65–134 (1956).

KEILIN, D., *The History of Cell Respiration and Cytochromes,* Cambridge University Press, Cambridge, 1966.

LEMBERG, M. R., Cytochrome Oxidase, *Physiol. Rev.* **49**, 48–121 (1969).

WAINIO, W. W., *The Mammalian Mitochondrial Respiratory Chain,* Academic Press, New York, 1970.

CHAPTER 5

KALCKAR, H. M., *Biological Phosphorylations, Development of Concepts,* Prentice-Hall, Englewood Cliffs, New Jersey, 1969.

KLINGENBERG, M., Metabolite Transport in Mitochondria: An Example for Intracellular Membrane Function, *in* P. N. Campbell and F. Dickens (eds.), *Essays in Biochemistry,* Vol. 6, Academic Press, New York, 1970.

LARDY, H. A., and FERGUSON, S. M., Oxidative Phosphorylation in Mitochondria, *Ann. Rev. Biochem,* **38**, 991–1034, (1969).

LEHINGER, A. L., *Bioenergetics,* W. A. Benjamin, New York, 1965.

RACKER, E., The Two Faces of the Inner Mitochondrial Membrane, *in* P. N. Campbell and F. Dickens (eds.), *Essays in Biochemistry,* Vol. 6, Academic Press, New York, 1970.

CHAPTER 6

GREEN, D. E., *Energy and the Mitochondria,* Academic Press, New York, 1970.

LEHNINGER, A. L., *The Mitochondrion: Molecular Basis of Structure and Function,* W. A. Benjamin, New York, 1964.

RACKER, E., (ed.), *Membranes of Mitochondria and Chloroplasts,* Van Nostrand-Rheinhold Co., New York, 1969.

CHAPTER 7

BALL, E. G., Regulation of Fatty Acid Synthesis in Adipose Tissue, *Advan. Enzyme, Regulation* **4**, 3–8 (1966).

FISCHER, E. H., POCKER, A., and SAARI, J. C., The Structure, Function and Control of Glycogen Phosphorylase, *in* P. N. Campbell and F. Dickens (eds.), *Essays in Biochemistry,* Vol. 6, Academic Press, New York, 1970.

WHELAN, W. J., and CAMERON, M. P., (eds.), *Control of Glycogen Metabolism,* Little, Brown and Co., Boston, 1964.

CHAPTER 8

ACHESON, G. H., (ed.), *Second Symposium on Catecholamines,* Section II, Metabolic Effects of Catecholamines, pp. 145–314, Williams & Wilkins Co., Baltimore, 1966.

ROBISON, G. A., BUTCHER, R. W., and SUTHERLAND, E. W., *Cyclic AMP,* Academic Press, New York, 1971.

SCRUTTON, M. C., and UTTER, M. F., The Regulation of Glycolysis and Gluconeogenesis in Animal Tissues, *Ann. Rev. Biochem.* **37**, 249–296 (1968).

INDEX